Daphna Weinshall, Jörn Anemüller, and Luc van Gool (Eds.)

Detection and Identification of Rare Audiovisual Cues

Studies in Computational Intelligence, Volume 384

Editor-in-Chief

Prof. Janusz Kacprzyk
Systems Research Institute
Polish Academy of Sciences
ul. Newelska 6
01-447 Warsaw
Poland
E-mail: kacprzyk@ibspan.waw.pl

Daphna Weinshall, Jörn Anemüller,
and Luc van Gool (Eds.)

Detection and Identification of Rare Audiovisual Cues

 Springer

Editors

Prof. Daphna Weinshall
Hebrew University of Jerusalem
School of Computer Science and Engineering
Jerusalem 91904
Israel
E-mail: daphna@cs.huji.ac.il

Prof. Jörn Anemüller
Carl von Ossietzky University
Medical Physics Section
Institute of Physics
26111 Oldenburg
Germany
E-mail: joern.anemueller@uni-oldenburg.de

Prof. Luc van Gool
K.U. Leuven
Department Electrical Engineering-ESAT
PSI-VISICS
Kasteelpark Arenberg 10 - bus 2441
3001 Heverlee
Belgium
E-mail: Luc.vanGool@esat.kuleuven.be
and
ETH-Zentrum
Computer Vision Laboratory
Sternwartstrasse 7
CH - 8092 Zürich
Switzerland
E-mail: vangool@vision.ee.ethz.ch

ISBN 978-3-642-24033-1   e-ISBN 978-3-642-24034-8

DOI 10.1007/978-3-642-24034-8

Studies in Computational Intelligence   ISSN 1860-949X

Library of Congress Control Number: 2011938287

© 2012 Springer-Verlag Berlin Heidelberg

Typeset & Cover Design: Scientific Publishing Services Pvt. Ltd., Chennai, India.

Printed on acid-free paper

9 8 7 6 5 4 3 2 1

springer.com

Introduction

Machine learning builds models of the world using training data from the application domain and prior knowledge about the problem. The models are later applied to future data in order to estimate the current state of the world. An implied assumption is that the future is stochastically similar to the past. The approach fails when the system encounters situations that are not anticipated from the past experience. In contrast, successful natural organisms identify new unanticipated stimuli and situations and frequently generate appropriate responses.

The observation described above led to the initiation of the DIRAC EC project in 2006. In 2010 a workshop was held, aimed at bringing together researchers and students from different disciplines in order to present and discuss new approaches for identifying and reacting to unexpected events in information-rich environments. This book includes a summary of the achievements of the DIRAC project in chapter 1, and a collection of the papers presented in this workshop in the remaining parts.

Specifically, in chapter 1 we define the new notion of *incongruent events*, aimed to capture "interesting" anomalous events. In part (ii) we present papers that describe how this conceptual approach can be turned into practical algorithms in different domains. Thus chapter (2) describes the detection of novel auditory events, chapter (3) describes the detection of novel visual events, chapter (4) describes the detection of out of vocabulary words in speech, and chapter (5) describes the detection of novel audio-visual events when the audio and visual cue disagree. Chapter (6) describes the data collected within the DIRAC project to test the approach and the different algorithms.

The chapters in part (iii) provide alternative frameworks for the identification and detection of interesting anomalous events. Both chapters deal with the detection of interesting events in video. Chapter (7) seeks the recognition of abnormal events, defining anomalies via trajectory analysis. Chapter (8) starts from the notion of Bayesian surprise, and develops a framework to detect surprising events based on the Latent Dirichlet Allocation model.

In part (iv) we focus on the question of what to do next? How to deal with those anomalous events we have detected? Thus chapter (9) talks about transfer learning from one rule-governed structure to another. Chapter (10) describes a retraining mechanism which learns new models once incongruence has been detected. Chapter (11) describes how to use a transfer learning algorithm in order to update internal models with only small training samples.

In part (v) we go back to the initial motivating observation, asking ourselves how biological systems detect and deal with unexpected incongruent event. In chapter (12) we present evidence that perception relies on existing knowledge as much as it does on incoming information. In chapter (13) we study mechanisms which allow the biological system to recalibrate itself to audio-visual temporal asynchronies. In chapter (14) we present a comparative study between biological and engineering systems in incongruence detection, in the context of locomotion detection.

Contents

Part III: Alternative Frameworks to Detect Meaningful Novel Events

Part IV: Dealing with Meaningful Novel Events, What to Do after Detection

Part V: How Biological Systems Deal with Novel and Incongruent Events

Part I
The DIRAC Project

DIRAC: Detection and Identification of Rare Audio-Visual Events

Jörn Anemüller, Barbara Caputo, Hynek Hermansky, Frank W. Ohl,
Tomas Pajdla, Misha Pavel, Luc van Gool, Rufin Vogels, Stefan Wabnik,
and Daphna Weinshall

Abstract. The DIRAC project was an integrated project that was carried out between January 1st 2006 and December 31st 2010. It was funded by the European Commission within the Sixth Framework Research Programme (FP6) under contract number IST-027787. Ten partners joined forces to investigate the concept of rare events in machine and cognitive systems, and developed multi-modal technology to identify such events and deal with them in audio-visual applications.

Jörn Anemüller
Carl von Ossietzky University Oldenburg, Germany
e-mail: joern.anemueller@uni-oldenburg.de

Barbara Caputo
Fondation de l'Institut Dalle Molle d'Intelligence Artificielle Perceptive, Martigny
Switzerland

Hynek Hermansky
Brno University of Technology, Czech Republic

Frank W. Ohl
Leibniz Institut für Neurobiologie, Magdeburg Germany

Tomas Pajdla
Czech Technical University in Prague, Czech Republic

Misha Pavel
Oregon Health and Science University, Portland USA

Luc van Gool
Eidgenössische Technische Hochschule Zürich, Switzerland

Rufin Vogels
Katholieke Universiteit Leuven, Belgium

Stefan Wabnik
Fraunhofer Institut Digitale Medientechnologie, Oldenburg Germany

Daphna Weinshall
University of Jerusalem, Jerusalem Israel

D. Weinshall, J. Anemüller, and L. van Gool (Eds.): DIRAC, SCI 384, pp. 3–35.
springerlink.com   © Springer-Verlag Berlin Heidelberg 2012

This document summarizes the project and its achievements. In Section 2 we present the research and engineering problem that the project set out to tackle, and discuss why we believe that advance made on solving these problems will get us closer to achieving the general objective of building artificial cognitive system with cognitive capabilities. We describe the approach taken to solving the problem, detailing the theoretical framework we came up with. We further describe how the inter-disciplinary nature of our research and evidence collected from biological and cognitive systems gave us the necessary insights and support for the proposed approach. In Section 3 we describe our efforts towards system design that follow the principles identified in our theoretical investigation. In Section 4 we describe a variety of algorithms we have developed in the context of different applications, to implement the theoretical framework described in Section 2. In Section 5 we describe algorithmic progress on a variety of questions that concern the learning of those rare events as defined in our Section 2. Finally, in Section 6 we describe our application scenarios, an integrated test-bed developed to test our algorithms in an integrated way.

1 Introduction

The DIRAC project was about rare events. Why rare events? We motivate our question in Section 2.1. In Section 2.2 we propose a theoretical framework which answers a related question, what are rare events? Finally, In Section 2.3 we discuss evidence from biological and cognitive systems, which support our choice of question and our proposed solution.

1.1 Motivation: The Problem

Since the introduction of von Neumann-like computing machines it was gradually becoming clear to most developers and users of information technologies that such machines, while increasing their computing power following the exponential Moore's law for many decades, are still failing behind biology on some seemingly very basic tasks. We have therefore assembled a group of likely-minded researchers, combining expertise in physiology of mammalian auditory and visual cortex and in audio/visual recognition engineering with the goal do discover what might be some of the fundamental issues that are preventing machines from being more effective on most cognitive tasks. All partners in the project agreed that understanding biological cognitive functions and emulating the selected ones in information extraction by machine is a way to achieving more efficient technology.

Among the fundamental machine weaknesses, we have identified one as particularly annoying: Machines work relatively well as long as the there is enough training data that describes well the information-carrying items that the machine needs to recognize. The machine fails when it encounters an unlikely or entirely unexpected item, typically recognizing it as one of the items from those that are expected. In contrast, it seems well founded that the unexpected items are the ones that get immediate attention when encountered by most biological systems.

Addressing this fundamental discrepancy between the machine and the biological organisms is bound to produce some interesting challenges.

We therefore specified our aims, to design and develop an environment-adaptive autonomous active cognitive system that will detect and identify rare events from the information derived by multiple, active information-seeking sensors. The system should probe for relevant cues, should autonomously adapt to new and changing environments, and should reliably discard non-informative data.

Our first challenge was to define what we mean by rare events. One possibility was to focus on being able to recognize items that do not occur in the environment of a given sensor too often – this meant being able to deal with very small amounts of training data from the given modality. This is closely related to the problem of outlier detection. Alternatively, we may want to think about incongruity between modalities, or some incongruity between an event and the context in which it occurs.

During this first year of the project, we reached an agreement that the project is about detecting and identifying events that are unexpected from the system point of view, i.e. the events or items that have low prior probability given the previous experience of the system. We then came up with a principled way to go about this question, by comparing predictions made by several (at least two) systems with different degrees of prior experience.

In our final approach, described formally in Section 2.2, we have proposed and investigated a general, biologically-consistent, strategy to detect the unexpected low-prior probability events. Subsequently we built several applications that followed this principle. The strategy relies on multiple information processing streams with different levels of prior constraints. This allows for the detection of instants where the incoming sensory data as evaluated by the more general model do not agree with the predictions implied by the more specific model. This strategy has been applied in detection of new faces for which a face classifier was not trained, new words that are not in the build-in dictionary of a speech recognizing machine and of new patterns of motion in video that do not obey constraints imposed by a model of expected human motions.

1.2 Definition and Theory

Our definition of rare events is based on the observation that there are many different reasons why some stimuli could appear rare or novel. Here we focus on those unexpected events, which are defined by the incongruence between a prediction induced by prior experience (training data) and the evidence provided by the sensory data. To identify an item as incongruent, we use two parallel classifiers. One of them is strongly constrained by specific knowledge (either prior knowledge or data-derived during training), the other classifier is more general and less constrained. Both classifiers are assumed to yield class-posterior probabilities in response to a particular input signal. A sufficiently large discrepancy between posterior probabilities induced by input data in the two classifiers is taken as evidence that an item is incongruent.

Thus, in comparison with most existing work on novelty detection, one new and important characteristic of our approach is that we look for a level of description where the novel event is sufficiently probable. Rather than simply respond to an event which is rejected by all classifiers, which often requires no special attention (as in pure noise), we construct and exploit a hierarchy of representations. We attend to those events which are recognized (or accepted) at some more abstract levels of description in the hierarchy, while being rejected by the classifiers at the more specific levels.

More specifically, we assume that the set of labels (or concepts) represents the knowledge base about the stimuli domain, which is either given (by a teacher) or learned. In cognitive systems such knowledge is hardly ever a set; often, in fact, labels are given (or can be thought of) as a hierarchy. In general, a hierarchy can be represented by a directed graph, where each label (a set of objects) corresponds to a single node in the graph. A directed edge exists from label (concept) a to b, if a (the more specific concept) corresponds to a smaller set of events or objects in the world, which is contained in the set of events or objects corresponding to label b, i.e., $a \subset b$. In this way the edges represent a partial order defined over the set of labels or concepts.

Because the graph is directed, it defines for each concept a two distinct sets of concepts (parent-child) related to it: *disjunctive concepts* which are smaller (subsets) according to the partial order, i.e. they are linked to node a by incoming edges converging on a; and *conjunctive concepts* which are larger (supersets) according to the partial order, i.e. they are linked to node a by outgoing edges diverging from a. If the DAG of partial order is a tree, only one of these sets is non trivial (larger than 1).

We consider two possible tree-like hierarchies, which correspond to two intuitive cases:

Conjunctive Hierarchy: Modeling part membership, as in biological taxonomy or speech. For example, eyes, ears, and nose combine to form a head; head, legs and tail combine to form a dog; and sequences of phoneme constitute words and utterances. In this case, each node has a single parent and possibly many children.

Disjunctive Hierarchy: Modeling class membership, as in human categorization – where objects can be classified at different levels of generality, from subordinate categories (most specific level), to basic level (intermediate level), to super-ordinate categories (most general level). For example, a Beagle (sub-ordinate category) is also a dog (basic level category), and it is also an animal (super-ordinate category), see right panel of Fig. 1. In this case, each node has a single child and possibly many parents.

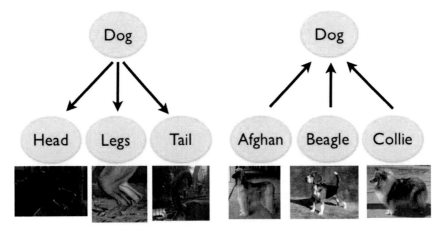

Fig. 1 Examples. Left: Conjunctive hierarchy, the concept of a dog requires the conjunction of parts, including head, legs and tail. Right: Disjunctive hierarchy, the concept of a dog is defined as the disjunction of more specific concepts, including Afghan, Beagle and Collie.

Multiple Probabilistic Models for Each Concept

For each node a, define A_s - the set of disjunctive concepts, corresponding to all nodes more specific (smaller) than a in accordance with the given partial order. Similarly, define A_g - the set of conjunctive concepts, corresponding to all nodes more general (larger) than a in accordance with the given partial order.

For each node a and training data T , we hypothesize 3 probabilistic models which are derived from T in different ways, in order to determine whether a new data point X can be described by concept a:

- $Q_a(X)$: a probabilistic model of class a, derived from training data T unconstrained by the partial order relations in the graph.
- $Q^s_a(X)$: a probabilistic model of class a which is based on the probability of concepts in A_s, assuming their independence of each other. Typically, the model incorporates a simple disjunctive relation between concepts in A_s.
- $Q^s_a(X)$: a probabilistic model of class a which is based on the probability of concepts in A_g, assuming their independence of each other. Here the model typically incorporates a simple conjunctive relation between concepts in A_g.

Definition of Incongruent Rare Events

In general, we expect the different models to provide roughly the same probabilistic estimate for the presence of concept a in data X. A mismatch between the predictions of the different models may indicate that something new and interesting had been observed, unpredicted by the existing knowledge of the system. In particular, we are interested in the following discrepancy:

Definition: Observation X is incongruent if there exists a concept a such that $Q^s_a(X) >> Q_a(X)$ or $Q_a(X) >> Q^s_a(X)$. In other words, observation X is incongruent if a discrepancy exists between the inference of two classifiers, where the more general classifier is much more confident in the existence of the object than the more specific classifier.

Classifiers come in different form: they may accept or reject, they may generate a (possibly probabilistic) hypothesis, or they may choose an action. For binary classifiers that either accept or reject, the definition above implies one of two mutually exclusive cases: either the classifier based on the more general descriptions from level g accepts X while the direct classier rejects it, or the direct classifier accepts X while the classifier based on the more specific descriptions from level s rejects it. In either case, the concept receives high probability at some more general level (according to the partial order), but much lower probability when relying only on some more specific level.

1.3 Evidence from Biological Systems

In parallel to the theoretical investigation described above, we have investigated mechanisms underlying the detection of rare incongruous events in various ways. Thus, using a combination of neurophysiological and behavioral experiments, we have rigorously demonstrated that the top-down mechanisms of rare-event processing we are interested in must not be confused with "mere novelty detection". In Section 5.3.we focus on a particular top-down mechanism, helping biological systems to respond meaningfully to unexpected rare events.

Traditionally, neuronal mechanisms underlying novelty detection are hypothesized to be reflected by the increased neuronal responses to deviant stimuli presented in the context of repeated, so called "standard", stimuli. This phenomenon is fundamentally a consequence of another well-known property of neuronal responses in sensory systems, namely stimulus-specific adaptation, i.e. the decaying neuronal response strengths with repeated presentation of identical stimuli. However, within DIRAC, we have emphasized that such bottom-up mechanisms of novelty detection can be conceptually and experimentally dissociated from top-down mechanisms of incongruence detection. Specifically, in one experiment we have been able to partial out the effects contributed by deviation of an unexpected stimulus with respect to its probability of occurrence (rareness) and contributed by deviating from the semantic context in which a stimulus can be expected. Moreover, the design of that experiment allowed us to make this comparison on the basis of neuronal responses to the exact same physical stimulus, which is a conceptual advantage to previous designs in this research area.

Specifically, in this experiment two groups of rodents (gerbils, *Meriones unguiculatus*) were trained to categorize 4 vowels from human speech in two orthogonally different ways. The two orthogonally different ways of forming a categorization boundary in the stimulus feature space allowed establishing two semantic contexts for identical features.

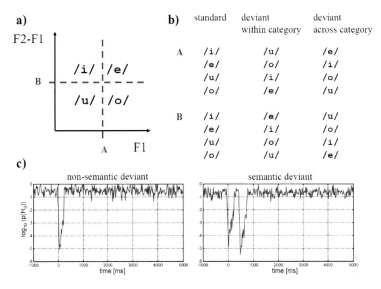

Fig. 2 Panel a) shows the positions of the 4 vowels in the feature space spanned by the first formant (F1) and the spectral distance between the first and second formant. This space basically conforms to the classical vowel feature space, but was shown to be physiologically realized in mammalian auditory cortex. One group of gerbils was trained to categorize the 4 vowels according to category boundary A, while the other group according to boundary B. The categorization training was realized as a standard active foot-shock avoidance Go-NoGo procedure, i.e. one category of stimuli required a Go response (jumping across a hurdle in a 2-compartment box) to avoid a foot-shock, the other category required a NoGo response (remaining in the current compartment of the box) to avoid the foot-shock. Thereby stimuli from the two categories acquire different meaning with respect to the appropriate behavior in the experiment. After training, classical novelty-detection ("odd-ball") experiments were conducted with both groups by presenting one vowel repeatedly as the standard stimulus and a second vowel as the infrequent deviant. Note that, given the previous vowel-categorization training, this second vowel could be selected either to be a member of the same meaning category as the standard stimulus or to be a member of the opposite category (associated with the opposite meaning with respect to required Go or NoGo behavior). Panel b) shows the different combinations of standard and deviants used in the post-categorization tests. Note that in the two orthogonal classification schemes (A and B), physically identical stimuli played the role of within-category deviants and across-category deviants, respectively.

For neurophysiological analysis we recorded multichannel electrocorticograms from auditory cortex as these signals have been demonstrated to provide physiological correlates of category formation during learning. Spatial patterns of electrocorticograms were used to classify vowel identity, and classification performance was analyzed in consecutive time bins of 120 ms (stepped in 20-ms steps) by comparing the number of correct classifications across all experimental trials with the expected number of correct classifications by chance (for details of the method see [8]). For each empirically found number of correct classifications, panel c) shows the probability of observing this number of correct classifications by chance (null hypothesis), separately for deviants being a member of the same meaning class (non-semantic deviants) and for deviants being a member of the opposite meaning class (semantic deviants). Significantly ($p < 10^{-3}$) different electrocorticogram patterns were found for both types of deviants at stimulus onset, but only for semantic deviants during an additional time window of 300 to 500 ms post stimulus onset. This latter result seems to indicate the existence of a physiological process mediating the detection of a top-down (meaning) incongruence, well separable from the bottom-up incongruence with respect to mere presentation statistics. More generally this phenomenon is in accordance with the general framework of incongruous-event detection in DIRAC as relying on detecting a mismatch between classifiers on a general level (vowel detectors) and a more specific level (vowels of class A or B).

A hierarchy of representations (e.g. from general to specific) is a fundamental assumption of the DIRAC framework. To test whether there is a hierarchical representation of animate and inanimate objects in the cortex, we have performed a monkey fMRI study in which we presented images of monkey bodies, human bodies, monkey faces, human faces, man-made objects, fruits, sculptures, mammals and birds. We have also presented the same stimuli to humans in a human fMRI study. We have run 3 monkeys using both block and event-related designs. We have found patches of regions in the macaque Superior Temporal Sulcus (STS), coding for monkey bodies and animals (and for faces). The data provides evidence – at the fMRI macroscale – of a hierarchical representation of objects in the monkey and human brain. First, multivoxel pattern analysis of the activation patterns in two STS regions of interests showed that a greater selectivity in the anterior compared to the posterior STS. Second, the visual cortical regions in humans were activated by bodies of humans, monkeys, birds and mammals, while regions outside the visual cortex were specifically activated for images of human faces and human bodies, indicating again hierarchical processing: from general to specific, in agreement with the DIRAC framework.

2 System Design

In this chapter we describe our efforts towards system building, including work on visual sensors and the representation of visual data (Section 3.1), auditory tasks and the representation of auditory data (Section 3.2), and our own mobile system – AWEAR II (Section 3.3).

2.1 Visual Sensors and Representation

Inspired by human vision, we employed omni-directional cameras with 180 degree field of view. Image calibration and rectification based on calibration patterns as well as on matching images of unknown scenes have been implemented, tested and integrated into the processing of images from the mobile as well as static AWEAR platforms.

Feature extraction (such as SIFT, SURF, and MSER) has been tuned to omnidirectional images and extended to time-space domain in order to capture the geometry as well as motion of features in images and sequences. The first fully scale-invariant spatio-temporal feature detector that is fast enough for video processing or for obeying timing constraints, has been designed and successfully used for inter-video matching and action recognition.

Feature extraction and image matching have been verified by developing video processing platforms allowing to track cameras and localize the observer w.r.t. to an unknown scene. Thus, in combination with the new direction-of-sound-arrival detector, a new audio-visual sensor allowing to sense images and sound in a coherent observer-centered spatial relationship has been developed. The audio-visual sensor is able to deliver measurements in a static as well as in dynamic setup. The ability to register acoustic and visual sensing in a common (static as well as moving) coordinate system extended the state of the art in audio-visual processing and made it possible to detect incongruence between audio and video streams. It became the basis for incongruence detection demonstrations.

Motivated by far reaching applications in detection of rare events in surveillance and video processing, computational tracking and human action recognition schemes were developed and implemented to support incongruence detection for human actions at different semantic levels. HOG based human visual detection was gradually expanded by developing a number of specialized visual detectors for lower and upper human body parts and for human appearance correlated with different activities. The machine learning approach was used to extract relevant features of short activities and a method for visual-activity vocabulary construction has been designed. This was a key element for building tracker-trees for human activity detection and interpretation. Rather than being designed on the basis of pre-defined action classes, tracker-trees construction proved possible to automatically pick up the different types of actions present in the data, and to derive a tracker tree from those action classifications. It also demonstrated that it is possible to learn activities from few training data for those patterns classified as abnormal by the self-learned tracker tree. This makes it possible to let tracker trees evolve over time and react to changing environment and the appearance of new incongruence.

2.2 Auditory Tasks and Representation

Reliable detection and representation of acoustic events forms a building block in constructing audio- and audio-visual classifiers that extract meaningful information from the environment. Thereby, they serve to identify rare events and subsequently aid their further identification and, in the case of their re-occurrence, adaptive learning of new object classes.

Tasks to be accomplished for representing the acoustic environment can broadly be divided into two groups. Identification of the class as an acoustic event or object pertains to links a physical sound pressure wave to discrete object categories, such as "speech", "car" or "dog". In contrast, spatial representation of an acoustic scene typically requires at least two measurements of the acoustic wave–taken at different locations–that are processed in order to identify the location of acoustic events, largely independent of the category of the constituting events.

Categorization of Acoustic Events

Relevant categories for acoustic objects are context and task dependent. Scenes that contain at least some non-speech objects demand categorization of the sources' identity, termed acoustic event detection. Speech recognition as a more specialized task necessitates the subdivision of the broad group of "speech" into categories relevant for ultimately recognizing an utterance's meaning. Here, the about 40 speech phonemes are the appropriate categories.

Both tasks have been investigated in DIRAC using several sets of acoustic features that are extracted from the sound pressure waveform. A focus has been on robust features that are to a high degree invariant under environmental changes such as room acoustics and background noise, leading to the analysis of the signals' modulation patterns, i.e., energy fluctuations across time in different spectral sub-bands.

The detection and discrimination of speech and non-speech objects builds on amplitude modulation spectrogram features (AMS). It represents a decomposition of the signal along the dimensions of acoustic frequency, modulation frequency and time, which is computed by a (modulation) spectral decomposition of subband spectral power time courses in overlapping temporal windows. The processing stages of the AMS computation are as follows. The signal decomposition with respect to acoustic frequency is computed by a short-term fast Fourier transformation (FFT with 32 ms Hann window, 4 ms shift, FFT length 256 samples, sampling rate 8 kHz), followed by squared magnitude computation, summation into rectangular, non-overlapping Bark bands and logarithmic amplitude compression. Within each spectral band, the modulation spectrum is obtained by applying another FFT (1000 ms Hann window, 500 ms shift, FFT length 250 samples) to the temporal trajectories of sub-band log energy. Outputs in the 0 Hz and 1 Hz modulation bands are influenced by DC components in the (log-energy) spectral domain and discarded as a means to reduce effects of channel noise (see also below). Finally, an envelope extraction and a further logarithmic compression are applied. By construction, the AMS features are approximately invariant to a time-domain signal convolution with short impulse responses such as microphone transfer functions and early reverberation effects. Using these features, classifiers for specific acoustic objects such as "door", "keyboard", "telephone" and "speech" are learned using Gaussian-kernel support vector machines (SVM) using a 1-vs-all multi-class training approach. [1,2]

Localization of Acoustic Events

The localization of sound sources in an acoustic scene is an important basis for the detection and identification of acoustic objects, albeit being somewhat orthogonal to the task of categorization since object category and object location are generally independent. Localization methods commonly employ more than one measurement channel, i.e., the spatial sound-field is sampled at several locations in space. Signals recorded at the different microphones differ in dependence on the room transfer functions from the sound sources to the different sensors, an effect that in a first approximation can be idealized as the time-delays with which the signal from a single source arrives at the different microphones. Cross-correlation measures and generalizations thereof (broadly termed "generalized cross-correlation", GCC) are employed to analyze these inter-channel differences. The ansatz developed here combines the phase-transform generalized cross-correlation measure (GCC-PHAT) with support vector classification. During training, a linear SVM is adapted to classify individual segments of the GCC-PHAT function as indicating presence (or absence) of a localized acoustic source. Thereby, a spatial map (parameterized by azimuth angle) is obtained that displays estimated directions of acoustic sources. The map shows the (possibly simultaneous) presence of sources, and in a post-processing step source probability estimates for each time-point and each azimuth direction are computed from the SVM confidence scores.

The individual acoustic categorization and localization methods form modules that are combined in subsequent steps, outlined below, in order to construct full audio-only and audio-visual systems for the detection of rare events.

2.3 AWEAR 2.0 System: Omni-Directional Audio-Visual Data Capture and Processing

To investigate the full scope of our approach, we designed an audio-visual back-pack system which collects data that could be taken by smaller, wearable sensing systems. The primary goal by the Dirac partners was to look into the type of imagery which industry considers particularly important: cases in surveillance where cameras cannot be considered static (a problem with even fixed cameras on poles), traffic safety applications where the sensors are car- or pedestrian-borne, video analysis for automated summarization and retrieval, etc. Most methods assume static cameras, which often come with assumptions like foreground results from background subtraction, smooth motions, etc. As soon as such conditions break down, industry often finds itself without effective methods. Based on discussions with the reviewers, it was nonetheless decided that Dirac would focus on indoor applications, like the independent living one, where the AWEAR system still proved useful, but simply as a static apparatus.

Sensor-wise, the AWEAR system is equipped with two high resolution cameras with fish-eye lenses, as well as a Firewire audio capturing device with four microphones, two heading forward and two backward. A total of three computers (two for video, one for audio, the latter also acting as the controller of the entire system) process the incoming multi-modal data streams, powered by a battery pack that can sustain the system for up to 3 hours. All components, along with further supporting mechanic hardware, are mounted on a rigid frame. The total weight is 20kg (10kg of that for batteries). The system is shown in Fig.3.

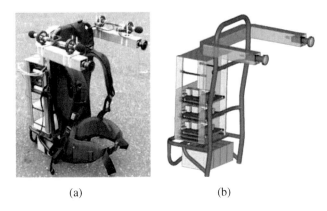

(a)   (b)

Fig. 3 (a) The AWEAR 2.0 system is comprised of 3 computers, 2 cameras heading forwards, and a Firewire audio capturing device with 4 microphones (2 heading forwards and 2 backwards). Thanks to 4 lead-gel batteries, the autonomy is about 3 hours. (b) AutoCAD model of the system.

The main benefits of the system are in particular (i) high resolution stereo, (ii) large field of view, and (iii) synchronization with multichannel high quality audio, and (iv) a wearable system. When searching for similar devices, one cannot find this combination.

Since not all the parameters were clearly known at design time, a modular design has been chosen. The platform can be extended to accommodate for four instead of two cameras or have the cameras replaced with faster ones capturing at double frame rate without having to modify the computing platform itself. Furthermore, up to four additional microphones can be added by just plugging them in. The computing platform has a margin in both bandwidth and processing power.

We used Ubuntu 8.10 as the operating system and several applications for video and audio capture. Video is captured in RAW (bayered) format into streams of 1000 files each, audio is saved as a 5-channel file, with the fifth channel containing the trigger pulses for video-audio synchronization.

For a system aimed at cognitive support, fish-eye lenses are very helpful due to their extended field of view. On the other hand, they required several dedicated steps for the data processing, going from calibration up to object class detection. Due to aberrations dependent on manufacturing and mounting, it is necessary to calibrate both lenses independently. For calibration, the entire field of view should be covered by a calibration target, rendering standard planar calibration targets unusable. We thus used a cube for that step. In order to find the transformation between the left and the right camera, we recorded a short sequence of 808 frames while walking in a room and then recovered the epipolar geometry as follows. Similarly, we developed debayering, geometrically rectifying, and several projection model cutout, structure-from-motion, image stabilization, and object class detection modules for AWEAR. For instance, to generate images more suitable for object recognition while keeping the full field of view, we used non-central cylindrical projection.

3 The Detection of Rare Events – Algorithms

We designed a number of application-specific algorithms, which implement the theoretical framework described in Section 2.2 and adapt the theory to the specific application. Three application domains are described below.

3.1 Visual and Audio Object Recognition

We adopted the framework described above to the problem of novel class detection, when given a **Disjunctive Hierarchy**. We assume a rich hierarchy, with non trivial (i.e. of size larger than 1) sets of *disjunctive concepts*, see right panel of Fig. 1. This assumption allows for the use of discriminative classifiers. We developed two applications: an algorithm to detect a new visual object, and an algorithm to detect a new auditory object.

Recall that in a disjunctive hierarchy we have two classifiers for each label or concept: the more general classifier $Q_{concept}$, and the specific disjunctive classifier $Q^s_{concept}$. The assumed classification scenario is multiclass, where several classes are already known.

Novel Sub-classes of Visual Objects

In order to identify novel classes, our algorithm detects a discrepancy between $Q_{concept}$ and $Q^s_{concept}$. The classifier $Q_{concept}$ is trained in the usual way using all the examples of the object, while the specific classifier $Q^s_{concept}$ is trained to discriminatively distinguish between the concepts in the set of disjunctive concepts of the object. Our approach is general in the sense that it does not depend on the specifics of the underlying object class recognition algorithm. We tested the algorithm experimentally on two sets of visual objects: a facial data set where the problem is reduced to face verification, and the set of motorbikes from the Caltech256 benchmark dataset.

Fig. 4 shows classification rates for the different types of test samples: Known - new samples from all known classes during the training phase; Unknown - samples from the unknown (novel) class which belong to the same General level as the Known classes but have been left out during training; Background - samples not belonging to the general level which were used as negative examples during the General level classifier training phase; and Unseen - samples of objects from classes not seen during the training phase, neither as positive nor as negative examples. The three possible types of classification are: Known - samples classified as belonging to one of the known classes; Unknown - samples classified as belonging to the unknown class; and Background - samples rejected by the General level classifier.

The results in Fig. 4 show the desired effects: each set of samples - Known, Unknown and Background, has the highest rate of correct classification in its own category. As desired, we also see similar recognition rates (or high acceptance rates) of the Known and Unknown classes by the general level classifier, indicating that both are regarded as similarly belonging to the same general level. Finally, samples from the Unseen set are rejected correctly by the general level classifier.

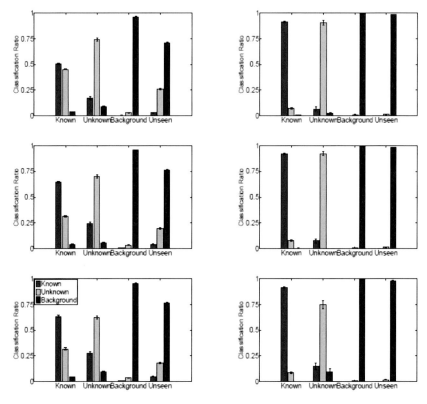

Fig. 4 Classification ratios for 4 groups of samples: Known Classes, Unknown Class, Background and sample of unseen classes. Bars corresponding to the three possible classification rates are shown: left bar shows the known classification rate, middle bar shows the unknown classification rate, and right bar shows the background classification rate (rejection by the general level classifier). The panels on the left correspond to the Motorbikes general level class. The panels on the right are representative plots of the Faces general level class.

Novel Sub-classes of Auditory Objects

We used the same algorithm as used above with an application from the domain of audio object classification, in order to evaluate the proposed framework in a different modality under systematically controlled noise levels. Here, the task is to discriminate known from novel audio objects appearing in an ambient sound background of a typical office environment. Hence, the inputs fall into three broad groups: Pure background noise (ambient environmental sounds such as ventilation noise recorded in an office room) with no specific audio object; known audio object embedded in background noise at a certain signal-to-noise ratio (SNR); and novel audio object embedded in the background at some SNR. Four classes of objects were considered: door opening and closing, keyboard typing, telephone ringing and speech. The non-speech sounds and the noise background were re-corded on-site, speech was taken from the TIMIT database. The continuous audio signals were cut into one second long frames, on which the analysis described be-low was carried out. Like before, performance is evaluated in a leave-one-out pro-cedure, i.e., each of the office objects is defined as novel once and left out of the training set.

The resulting performance levels at equal error rate (EER) are displayed in Fig. 5. Here, the performance is bounded from above by the (arbitrary) choice of 5% false positive for the tuning of the general classifier. The results demonstrate that the detection of Unknown objects based on a hierarchy of classifiers is possible in the acoustic domain and its performance depends on the type of novel signal and SNR.

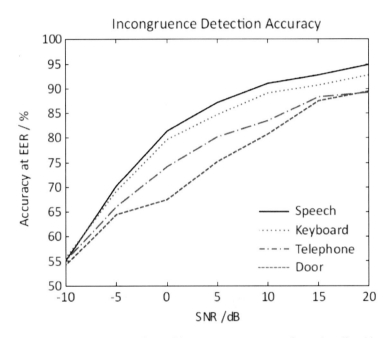

Fig. 5 Accuracy of novelty detection, with one curve per type of novel audio object (see legend). The accuracy is taken at the EER point (equal false alarm and miss rates). Below 10dB, the EER could not be determined. Note that the accuracy is bounded from above by the (arbitrary) choice of 5% false positive rate for the general classifier.

3.2 Out of Vocabulary Words in Speech Processing

Current large vocabulary continuous speech recognition systems (LVCSR) are customized to operate with a limited vocabulary on a restricted domain. As prior knowledge, text-derived language models (LM) and pronunciation lexicons are utilized, and are designed to cover the most frequent words and multi-grams.

Under real conditions, however, human speech can contain an unlimited amount proper names, foreign and invented words. Thus, unexpected lexical items are unavoidable.

If a word is missing in the dictionary (out-of-vocabulary - OOV), the probability of any word sequence containing this word according to such LM is zero. As a consequence, the corresponding speech will be mis-recognized - the OOVs are replaced by acoustically similar in-vocabulary (IV) words. The information contained in the OOVs is lost and cannot be recovered in later processing stages. Due to the contextual nature of the LM, also the surrounding words tend to be wrong. Since OOVs are rare, they usually do not have a large impact on the Word Error Rate (WER). However, information theory tells us that rare and unexpected events are likely to be information rich. Improving the machine ability to handle unexpected words would considerably increase the utility of speech recognition technology.

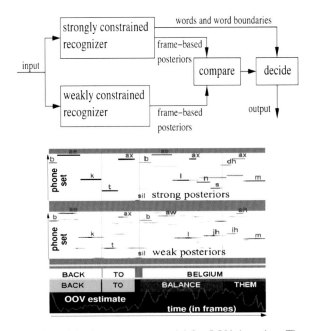

Fig. 6 Left: Application of the incongruence model for OOV detection. The generic model is the weakly constrained recognizer and the specific model is the strongly constrained recognizer. Right: Phone posteriors and OOV detection output using NN-based comparison for the OOV "Belgium".

Within the DIRAC project a novel technique for the detection of unexpected and rare words (especially OOV) in speech has been proposed and developed. The approach is based on the comparison of two phoneme posterior streams (Fig. 6) derived from the identical acoustic evidence while using two different sets of prior constraints - strongly constrained (LVCSR, word-based, with LM) and weakly constrained (only phones) [4,13,36]. We aim to detect both where the recognizer is unsure and where the recognizer is sure about the wrong thing. The mismatch between the two posterior streams can indicate an OOV, although the LVCSR itself is quite sure of its output.

The hierarchical rare events detection scheme has been shown to outperform related existing posterior based confidence measures (CM) when evaluated on a small vocabulary task, where the posterior estimates in the two channels were compared by evaluating the Kullback-Leibler divergence at each frame. After that, the use of frame-based, word- and phone- posterior probabilities ("posteriors") as CM was further investigated on a large vocabulary task (Wall Street Journal data) with reduced recognition vocabulary. The task was to classify each recognized word as either being OOV or IV, based on a word confidence score, obtained from averaging frame level CMs over the boundaries in the recognition output. With the introduction of a trained comparison of posterior streams (using a neural net – NN) and using new hierarchical techniques for estimating phoneme posteriors, significant improvement over state-of-the-art posterior-based CM was achieved.

After that, the NN based OOV word detection was applied to noisy, lower quality telephone speech (CallHome, Eval01, Fisher) to show the robustness of the approach. In addition, the classification performance improved by classifying the recognized word output using several classes (IV correct, IV incorrect, OOV, silence) [17].

The vocabulary used in speech databases usually consists of two types of words: firstly, a limited set of common words, shared across multiple documents (typically IV); secondly, a virtually unlimited set of rare words, which might only appear a few times in particular documents (mostly OOV). Even if these words do not have a big impact on the word error rate, they usually carry important information. OOVs which occur repeatedly often represent topic-specific terminology and named entities of a domain. Therefore, we further concentrated on the detection of frequently re-occurring unexpected words. To achieve this, we ran our OOV detection system on telephone calls and lectures centered around a certain topic [9].

For all further experiments, OOV detection should serve as an instrument to obtain descriptions of OOV words - putting special emphasis on repeatedly occurring OOVs. The task is now to detect the time span of the reference OOV word as completely and precisely as possible. Since the NN-based OOV detection system offers no description/boundaries of the OOV, we subsequently integrated a hybrid word/sub-word recognizer [29] into our system (Fig. 7), which lets us obtain boundaries and descriptions of OOVs words in an integrated way. The system is not required to substitute an OOV by some IV; it can fall back to its sub-word model and thus retrieve a lower-level description of the word in terms of sub-word units. Here the boundaries for OOV words were estimated more accurately than with the NN-based system.

The artificially high OOV rates due to the use of small decoding vocabularies were identified as a problem. Therefore, we finally changed to a much larger decoding vocabulary and a data set consisting of a collection of topic-specific TED talks, in which we find a reasonably high number of information-rich OOV words [16]. It represents a more realistic scenario, since only naturally appearing and harder to detect OOVs are targeted.

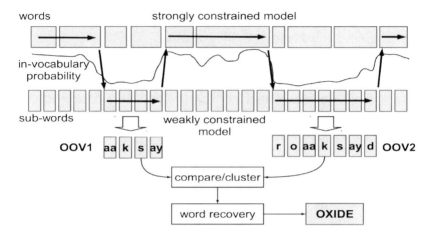

Fig. 7 Scheme of a hybrid word/sub-word model for OOV detection. The decoder has the freedom to compose the path (recognized word sequence) of words from either the specific word or the generic sub-word model. The detected sub-word sequences from the best path are taken as OOV candidates for similarity scoring and recovery.

To deal with rare and new words in ASR we proposed follow-up actions that can be taken after the detection of an OOV to analyze the newly discovered words and to recover from the mis-recognitions. The goal is to avoid mis-recognitions in the presence of rare words by designing a system that is open-vocabulary and that can learn with its usage. The hybrid word/sub-word recognizer solves the OOV localization and obtains its phonetic description – the detected phoneme sequence in the detected time span - in an integrated way [16].

Given the location and phonetic description of an OOV, one possible action is to recover the orthographic spelling of the OOV. We showed that OOV spelling recovery [16] can successfully recover many OOVs, lowering the word error rate and reducing the number of false OOV detections.

Aiming at topic-specific repeating OOVs, we introduced the task of similarity scoring and clustering of detected OOVs. A similarity measure [9] based on aligning the detected sub-word sequences was developed, which serves to identify similar candidates among all OOV detections. A new form of word alignment is introduced, based on aligning the OOV to sequences of IVs/other OOVs, which retrieves a higher-level description of the OOV, in the sense of word relations. (e.g. being a compounded word or a derivation of a known word).

Finally, we have investigated into novel language modeling techniques based on Neural networks and recurrent neural networks [18] that offer the possibility of implicit smoothing. In addition to their excellent performance compared to standard n-gram models, they are also suitable for ASR techniques coping with out-of-vocabulary words.

3.3 Biological Motion

In our third application, we developed a system for the detection of incongruent events that is based on the detection of activities, i.e. motion patterns. We compiled a set of motion data that contains walking and running patterns by several subjects, at different speeds. Subjects were placed on a conveyor belt, so that a motion capture system could capture the data. The system delivered the 3D coordinates of the body joints as they evolved over time. Moreover, normal cameras were also taking videos of the same actions, from which a series of silhouettes were obtained from 8 different viewpoints. The computational work then started from these silhouettes, whereas the neurophysiological work (described in the next section) started from different types of stick like figures, based on the motion captured data of exactly the same actions.

Of course, in order to tackle real-life problems, it was necessary to capture quite a broader range of activities. Thus we developed the so-called tracker trees. These are hierarchies of trackers, with a very generic blob tracker at the root node, and becoming more and more specific when moving to higher layers. For instance, a walking tracker has been trained, which would respond to the walking pattern of any subject. At the higher level, a layer of individual walking trackers is found, which respond to the particular gait of the corresponding people. In this way, the tracker tree can spot incongruent events of very different nature, e.g. a dog entering the apartment if there normally is none, a person falling, or another person than those known to the system entering the house. The system can also indicate what kind of incongruency occurred, at least qualitatively. This is important as the kind of action to take depends on the nature of the particular incongruency.

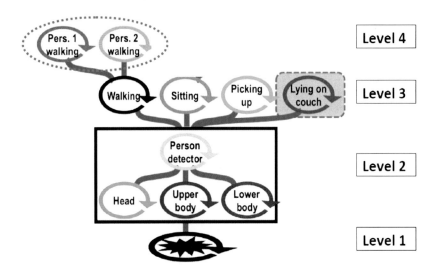

Fig. 8 Visualization of the tracker tree and its dependencies over multiple levels.

We subsequently added several upper body activities detectors to the basic tracker tree, including such actions as reading or drinking. These required a more involved analysis than the rest of the activities, as they are more difficult to distinguish from silhouettes. We used our own spatio-temporal features - 3D extensions of SURF features.

In a further move, we extended the tracker trees with self-learning capabilities. This is important as in practical situations extensive training may be overly expensive, e.g. if one would want to install such systems in the homes of many people. First, a dual hierarchy has been developed, to mirror the snapshot - motion duality found in the neurophysiological research [34],described in the next section. This type of tracker tree would automatically pick up new types of silhouettes or silhouette sequences. The response of this tracker tree has also been compared to neuronal responses to exactly the same dataset (see first paragraph). Strong qualitative similarities could be found.

We then set out to further improve the results and the flexibility of the self-learning tracker trees. A novel approach was introduced, where the hierarchy is discovered through Slow Feature Analysis. Once activity nodes have been built, these are described on the basis of PCA, to well capture the variability among activities of the same class. This type of self-learning tracker tree allows for other than fixed splits (e.g. binary), as was assumed with the initial type.

In parallel we invested resources in exploring applications other than independent living, especially during the final phase of the project. Several extensions have been demonstrated, including outdoor usage, surveillance application, automated detection of abnormal images captured by webcams, etc.

Moreover, in order to allow for self-adaptive tracker trees and automatically add extra nodes, we integrated it with the transfer learning method described below in Section 5.2. After the initial tracker tree has been trained, it would start to detect events that according to the training material are incongruent. Applying the Transfer Learning principles we could automatically add new nodes for the incongruent events. This then allows us to name them, and take appropriate action for each case. In this way a small number of training events sufficed to build the additional nodes.

3.4 Biological Motion as Perceived by Biological Systems

In order to detect an incongruent event it is essential that one has a model of congruent events. Thus the aim of this section is to understand how biological organisms, in particular primates, represent actions of people. The computational principles discovered in these neurophysiological studies suggested and supported the computational framework described above in Section 4.3 and were used to detect actions, in particular walking/running people.

In a first series of studies [34], we studied the spiking responses of single macaque temporal cortical (rostral Superior Temporal Sulcus (STS)) neurons to a parameterized set [3] of dynamic visual images of actions. We used arm actions like knocking, lifting and throwing and their morphs. The action images were rendered as stick figures. We found that as a population, the neuronal population represented the similarity among the different actions, as shown by a non-linear multidimensional scaling (ISOMAP) of the pairwise differences between the neural responses to the different stimuli. We were able to distinguish different kinds of neuronal selectivity. Firstly, neurons, mainly in the ventral bank of the rostral STS, responded as well to the action movies as to static snapshots of these movies. These neurons clearly responded to form information. Secondly, other neurons, mainly in the dorsal bank of the rostral STS, responded much less to static snapshots than to the action movies, thus responding to motion information. This dual processing, form and motion based, has been used to develop the computational framework for action detection described above in Section 4.3.

In the next series of studies, we employed stimuli that were based on motion-capture data of real human subjects that were walking or running at different, controlled speeds on a treadmill, as described above in Section 4.3.. In a first phase, we performed an extensive behavioral study of the perception of these biological motion displays in monkeys [33]. We trained 3 macaques in the discrimination of facing-direction (left versus right) and forward versus backward walking using the above discussed motion-capture-based locomotion displays in which the body features were represented by cylinder-like primitives. Discriminating forward versus backward locomotion requires motion information while the facing-direction/view task can be solved using motion and/or form. All monkeys required lengthy training to learn the forward-backward task, while the view task was learned more quickly. Once acquired, the discriminations were specific to walking and stimulus format but generalized across actors. Performance in the forward-backward task was highly susceptible to degradations of spatio-temporal stimulus coherence and motion information. Importantly, collaborative computational work [19] showed that the walking-running speed generalization in the forward-backward discrimination fitted the predictions made using the DIRAC computational architecture developed by ETH, thus supporting this architecture.

After the behavioral training, we conducted single cell recordings in the trained animals, examining the contribution of motion and form information to the selectivity for locomotion actions. We recorded in both dorsal and ventral banks of the rostral STS. The majority of the neurons were selective for facing direction, while a minority distinguished forward from backward walking. We employed Support Vector Machines classifiers to assess how well the population of recorded neurons could classify the different walking directions and forward from backward walking. Support vector machines using the temporal cortical population responses as input classified facing direction well, but forward and backward walking less so but still significantly better than chance. Classification performance for forward versus backward walking improved markedly when the within-action response modulation was considered, reflecting differences in momentary body poses within the locomotion sequences. Analysis of the responses to walking sequences wherein the start frame was varied across trials showed that some neurons also carried a snapshot sequence signal. Such sequence information [6,35] was present in neurons that responded to static snapshot presentations and in neurons that required motion [32]. In summary, our data suggest that most STS neurons predominantly signal momentary pose. In addition, some of these temporal cortical neurons, including those responding to static pose, are sensitive to pose sequence, which can contribute to the signaling of learned action sequences [7]. Both mechanisms, the pose mechanism and the pose-sequence mechanism, have been incorporated into the model described above in Section 4.3.

4 The Learning of Rare Events

Now that we have detected those rare events, comes the question of what to do with them: How do we bootstrap some representation, or classifier, for these novel and yet interesting events? This chapter describe various studies which address related questions, such as online learning (Section 5.1) and knowledge transfer seen from a computational (Section 5.2) and biological (section 5.3) points of view.

4.1 Online Learning

The capability to learn continuously over time, taking advantage of experience and adapting to changing situations and stimuli is a crucial component of autonomous cognitive systems. This is even more important when dealing with rare events, where learning must start from little available data. From an algorithmic point of view, this means developing algorithms able to build novel representations from a few labeled samples, which are made available one at the time. At the same time, it is desirable for the same algorithm to be able to cope with little incoming data (this is the case when learning a detected rare event), and be able to choose to update the representation of a known class for which large amounts of data are available. This calls for online learning algorithms able to cope with small amount of data as well as large amounts, without suffering from memory explosion, while maintaining high performance and low algorithm complexity.

In one of our attempts to address this issue, we developed an algorithm - Online Incremental SVM, where we proposed an online framework with a theoretically bounded size for the solution. [23,26]. The basic idea was to project the new incoming data on the space of the current solution, and to add it to the solution only if it was linearly independent. The drawback of the method was that all the incoming data had to be stored in order to have an exact solution. We applied the same principle of projecting the incoming data on the space of the current solution to the perceptron, an online algorithm with forgetting properties. The resulting method, which we called projectron, has bounded memory growth and low algorithmic complexity. [22,25]. Perceptron-like algorithms are known to provide lower performance than SVM-based method.

We then moved to extend these results to multi modal data. [12]. We developed a Multi Kernel Learning algorithm that is state of the art in terms of speed during training and test, and in terms of performance on several benchmark database. We have called the algorithm OBSCURE (for "Online-Batch Strongly Convex mUlti keRnel lEarning algorithm"). [21]. It has guaranteed fast convergence rate to the optimal solution. OBSCURE has training time that depends linearly on the number of training examples, with a convergence rate sub-linear in the number of features/kernels used. At the same time, it achieves state-of-the-art performance on standard benchmark databases. The algorithm is based on a stochastic sub-gradient descent algorithm in the primal objective formulation. Minimizing the primal objective function directly results in a convergence rate that is faster and provable, rather than optimizing the dual objective. Furthermore, we show that by optimizing the primal objective function directly, we can stop the algorithm after a few iterations, while still retaining a performance close to the optimal one.

4.2 Knowledge Transfer

How to exploit prior knowledge to learn a new concept when having only few data samples is a crucial component for being able to react upon the detection of an incongruent event. This can be declined in two settings: transfer learning across models build on the same input modalities, and transfer learning across models built over different modalities. In both cases, we addressed the problem of transfer learning within a discriminative framework.

When dealing with knowledge transfer across models built on the same modality, the basic intuition is that, if a system has already learned N categories, learning the N+1 should be easier, even from one or few training samples. We focused on three key issues for knowledge transfer: how to transfer, what to transfer and when to transfer. We proposed a discriminative method based on Least Square SVM (LS-SVM) (how to transfer) that learns the new class through adaptation. [30,31]. We define the prior knowledge as the hyperplanes of the classifiers of the N classes already learned (what to transfer). Hence knowledge transfer is equivalent to constraining the hyperplane of the N+1 new category to be close to those of a sub-set of the N classes. We learn the sub-set of classes from where to transfer, and how much to transfer from each of them, via an estimate of the Leave One Out (LOO) error on the training set. Determining how much to transfer helps avoiding negative transfer. Therefore, in case of non- informative prior knowledge, transfer might be disregarded completely (when to transfer).

We also investigated the possibility to develop an algorithm able to mimic the knowledge transfer across modalities that happens in biological systems. [24]. We made the working assumption that in a first stage the system have access to the audio-visual patterns on both modalities, and that the modalities are synchronous. Hence the system learns the mapping between each audio-visual couple of input data. The classifier is designed as follows: each modality is classified separately by a specific algorithm, and the outputs of these classifiers are then combined together to provide the final, multi-modal classification. When the classifier receives an input data where one of the two modalities is very noisy, or completely missing, the internal model generates a 'virtual input' that replaces the noisy/missing one, and incrementally updates its internal representation. Experiments on different multimodal settings show that our algorithm improves significantly its performance in the presence of missing data in one of the two modalities, therefore demonstrating the usefulness of transfer of knowledge across modalities. [5] Moreover the framework is deeply rooted on the theory of online learning that gives theoretical guarantees on the optimality of the approach.

4.3 Knowledge Transfer in Rodents

In this section we focus on our neuro-behavioral research in rodents demonstrating that trans-modal category transfer can be used as a mechanism to respond meaningfully to unexpected stimuli in a given sensory modality. Specifically we investigated whether and how, knowledge acquired during learning about the relevance of stimulus features in one sensory modality (here, audition) can be transferred to novel, unexpected stimuli of another sensory modality (here, vision).

We trained rodents (gerbils, Meriones unguiculatus) to associate a slow and a fast presentation rate of auditory tone pips with the Go response and NoGo response, respectively, in an active avoidance paradigm (shuttle box). After a predefined performance criterion was reached for the discrimination task in the auditory modality, a second training phase was initiated in which the sensory modality of the stimuli was changed from auditory to visual. For one animal group (congruent group) the contingency of the two stimulus presentation rates with the Go/Nogo-responses stayed the same irrespective of the modality of stimulation, for a second group (incongruent group) it was reversed across modalities.

After the modality switch, the congruent groups showed a higher acquisition rate of the conditioned responses than the incongruent groups indicating a cross-modal transfer of the rate-response association. During training, the electrocorticogram was recorded from two multitelectrode arrays chronically implanted onto the epidural surface of primary auditory and the visual cortex.

Cortical activity patterns from the ongoing electrocorticogram associated with the Go- and the NoGo stimuli were determined in the spatial distribution of signal power using a multivariate pattern classification procedure. For animals of the congruent group showing correct discrimination already during the first visual training sessions, we suspect that these individuals transferred the rate-response association learned during auditory training to the visual training. In these animals activity-patterns observed in the ongoing electrocorticogram of the auditory and the visual cortex were associated both with the auditory and the visual Go- and the NoGo-stimuli. We suggest that activity in both auditory and visual cortex was instrumental for achieving the cross-modal transfer of learned associations.

5 Application Scenario

The one goal of the DIRAC project was to establish the paradigm of incongruency as a novel method of information retrieval within a model hierarchy. Research was inspired not only by theoretical questions, but also by the urge to find physiological evidence for special forms of hierarchies and models. With the incongruency reasoning formulated, fostered and the principles tested with first experiments conducted by the partners, a growing need for experimental data to further investigate and evaluate the findings led to the aggregation of different databases within the project. Over time, databases were assembled and used by the partners to drive their development and do verify their findings, as described in Section 5.1. Evaluation is described in Sections 5.2 and 5.3.

5.1 Datasets

The following sub-sections give descriptions of the database of audio-visual recordings, the database of multichannel in-ear and behind the ear head related and bin-aural room impulse responses, the database of frequency modulated sweeps for STRF estimation, the database of OOV and OOL recordings and the database of motion captured actor walking on a treadmill.

Database of Audio-Visual Recordings

Within the DIRAC project, two application domains were defined for rare and incongruent event detection, namely the security and surveillance on the one hand, and in-home monitoring of elderly people on the other.

Based on these application domains, scenarios have been developed by the project partners to show the potential of the DIRAC theoretical framework and the techniques developed within the project, while attempting to address realistic and interesting situations. Each scenario in turn was recorded by different partners using professional audio and video recording hardware assembled into two recording platforms, the AWEAR II and the OHSU recording setup. During the project, the partners refined both the developed detectors and the scene descriptions. That – over time – formed the audio-visual database of the DIRAC project. Several hundred recordings were recorded in more than 50 recording sessions at different locations.

The raw recording data had to be pre-processed, including format changes, projections to correct lens distortion, synchronization between video and audio, and preview video with reduced resolution, prior to the application of the detectors and model by the partners. The data was categorized using a list of keywords and time stamps to help partners to identify and search for different human actions contained in each recording. For evaluation purposes, the audio and video data of the recordings were annotated on a frame by frame basis, giving the pixel position of different body parts of the actor, e.g. the pixel position of the head and the upper and lower body, or the speech/non-speech classification of the recorded audio signal as time positions.

Database of OOV and OOL Recordings

Two data sets of audio recordings have been produced. The first one is a set of utterances containing Out-Of-Vocabulary (OOV) words and non-speech sounds, and the second one contains English In-Language (IL) spontaneous speech featuring intermittent switches to a foreign language (Out-Of-Language – OOL).

Database of Multichannel In-Ear and Behind the Ear Head Related and Binaural Room Impulse Responses

An eight-channel database of head-related impulse responses (HRIR) and binaural room impulse responses (BRIR) was generated within the DIRAC project. The impulse responses (IR) were measured with three-channel behind-the-ear (BTE) hearing aids and an in-ear microphone at both ears of a human head and torso simulator. The scenes' natural acoustic background was also recorded in each of the real world environments for all eight channels. Overall, the present database allows for a realistic construction of simulated sound fields for hearing instrument research and, consequently, for a realistic evaluation of hearing instrument algorithms. The database is available online in the DIRAC project website and described in more detail in [14].

Database of Frequency Modulated Sweeps for STRF Estimation

The spectro-temporal receptive field (STRF) is a common way to describe which features are encoded by auditory and visual neurons. STRFs are estimated by relating stimuli, visual or auditory, to the evoked response ensemble of the neuron. Once an STRF has been estimated it can be used to predict the linear part of the response of the neuron to new stimuli.

A new class of stimuli was generated within the DIRAC project consisting of frequency modulated (FM) sweeps. FM sweeps are an alternative to dynamic moving ripples (DMR) that are often used for STRF estimation. Both stimulus classes are designed such that they sample a large portion of the neuron's input space while having very low autocorrelations which makes them suitable for STRF estimation. However, as there is one main feature to divide the sweep class into two, the up and the down sweep, this characteristic can be used to investigate context dependence of the STRF. We showed that neurons as well as local field potentials (LFPs) are sensitive to different sweep orientation [32]. Hence, one could estimate STRFs using a varying number of up and down sweeps to realize different context conditions. The database and example scripts are available online in the DIRAC project website.

It has been shown that there are shortcomings in the STRF and its estimates cannot account for arbitrary stimuli. The impact on the scientific community is a potentially better description of the STRF. Furthermore, it is possible to realize different context conditions which may yield new insights into encoding strategies on the level of single neurons. In terms of DIRAC it would give a direct comparison if a stimuli is standard or deviant.

Database of Motion Captured Actor Walking on a Treadmill

We investigated how temporal cortical neurons encode actions differing in direction, both forward versus backward as different facing directions. The stimuli in this study were locomotory actions, i.e. displacement of a human body, but whereby the translational component is removed, thus resembling an actor as if locomoting on a treadmill. This was done in order to more accurately pinpoint the neural code sub-serving walking direction (walking left- or rightward either in a forward or backward fashion), for which actual physical displacements of the body are erroneous in determining the neural code.

Computationally, coding between forward and backward versions differs with respect to which aspects of the actions are relevant compared to coding for different facing directions. All possible combinations amount to 16 actions in total. All stimuli have been made available as *.avi such that playback is not limited to certain devices.

5.2 Integrated Algorithms

For the evaluation process, both audio and visual detectors provided by different project partners were evaluated against the ground truth annotation of the different databases of the DIRAC project. The following subsections provide an overview of each detector.

Acoustic Object Detection – Speech/Non-Speech Discrimination

This detector aims to classify parts of the one-channel audio signal of a recorded scene as either speech or non-speech using a pre-trained model. The detector produces a binary label output every 500 milliseconds, indicating whether speech was detected or not.

The model used for this detection is based on amplitude modulation features coupled with a support vector machine classifier back-end. Features used for classification are modulation components of the signal extracted by computation of the amplitude modulation spectrogram. By construction, these features are largely invariant to spectral changes in the signal, thereby allowing for a separation of the modulation information from purely spectral information, which in turn is crucial when discriminating modulated sounds such as speech from stationary backgrounds [2]. The SVM back-end allows a very robust classification since it offers good generalization performance (see also Section 3.2).

Acoustic Localization Detector

The acoustic localization detector analyzes a 2-channel audio signal of a recorded scene. It aims to give directional information of every acoustic object it detects within every time frame of 80 milliseconds. The output for every time frame is a vector of yes/no information for all 61 non-overlapping segments between 0 and 180 degrees of arrival with respect to the stereo microphone basis. The detector is capable of detecting multiple acoustic objects within one time frame, but not capable to classify any localized acoustic object.

The detector uses a correlation-based feature front-end and a discriminative classification back-end to classify the location-dependent presence or absence of acoustic sources in a given time frame. The features are computed on the basis of the generalized cross correlation (GCC) function between two audio input signals. The GCC is an extension of the cross power spectral density function, which is given by the Fourier transform of the cross correlation. Support Vector Machines are employed to classify the presence or absence of a source at each angle. This approach enables the simultaneous localization of more than one sound source in each time-frame (see also Section 3.2).

Tracker Tree

The tracker tree processes video information on a frame by frame basis by utilizing multiple specialized models organized in a hierarchical way (see Fig. 8). Each model operates on a frame by frame basis and gives confidence measures for the action it was designed for, i.e. "sitting" for a sitting person, "walking" for a walking or standing person, and "picking" for a person picking up something [20].

Conversation Detector

The multi-modal Conversation Detector uses the DIRAC principle of incongruency detection to discriminate between normal conversation and unusual conversational behavior, e.g. a person talking to himself. The Detector operates on output data from three detectors presented in this document: the speech/non-speech classifier, the audio localizer, the acoustic localization detector, and the tracker tree's person detector. The output signals are combined into a DIRAC incongruence model instantiating a part-whole relationship (see also Section 2.2 for description of the part-whole relationship model). Within the Conversation Detector, three models on the general level (PT: person tracker, AL: audio localizer, SC: speech classifier), are combined to one conjoint model; one model on the specific level uses the fused data input of each model on the general level (see Fig. 9). The model on the specific level (Cf) utilizes a linear support vector machine and operates on the same (albeit fused) input data as the models on the general level. An incongruency is detected when the conjoint model accepts the input as conversation, whereas the specific model does not.

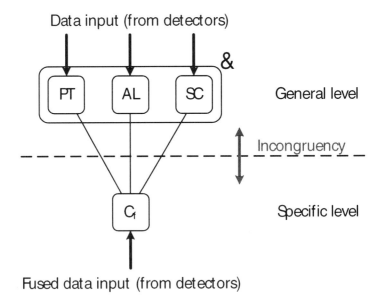

Fig. 9 DIRAC incongruence model of the Conversation Detector.

Combined Audio-Visual Incongruence Detector

Combined audio-visual incongruence detector is an example of a conjunctive hier-archy in audio-visual processing. Alternative detectors (i.e. discriminative classi-fiers) were used to model events in a hierarchical manner, see Figure 10. We con-centrate on the single audio-visual event of a human speaker in a scene and model it in two alternative ways. We assume a scene observed by a camera with wide view-field and two microphones. Visual processing detects the presence and posi-tion of a human. Sound processing detects the intensity of sound and its direction of arrival.

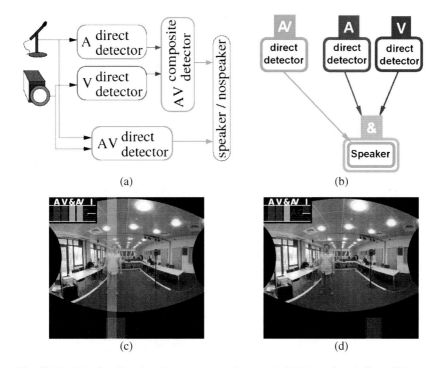

(a)   (b)

(c)   (d)

Fig. 10 Combined audio-visual incongruence detector. (a,b) Direct A and direct V detectors are combined into a composite A\V detector. Another direct A\V detector is run in parallel. Detectors are congruent (c) when they agree or incongruent (d) when the composite A\V detector is active while the direct A\V detector is passive.

The specific (direct) A\V is obtained by training a discriminative RBF SVM classifier on audio-visual features extracted from manually labelled training data of human speakers vs. background. It is evaluated on all spatial windows of meaningful size across the view-field, thus implicitly providing the positions of its decisions.

The general (composite) A\V detector is obtained by the conjunction of the di-rect visual detector and the direct audio detector. Unlike the direct A\V detector, it does not exploit the information about where the direct A and V detectors were active in the view-field. In effect, it looks whether they were active irrespectively of the position.

By construction, the composite A\V detector returns a positive outcome when observing a human body and human sound in different positions in the view-field. The direct A\V detector, on the other hand, is passive in this situation since it has been trained only on co-located human sound and visual examples. Thus, an incongruence appears. The appearance of such incongruence indicates a deficiency in the world model and can be used to initiate learning and updates of the model.

Combined Tracker Tree and Transfer Learning

We combined the tracker tree algorithm from for incongruent actions detection and the transfer learning in order to learn the new detected action. In this combination, the tracker tree detects an incongruent action, and asks for human annotation of few frames (from 1 to maximum 10) of it. These annotated frames are sent to the transfer learning algorithm, which learns the new action from the few annotated samples, exploiting the prior knowledge of the system. Note that the original tracker tree algorithm would need an average of 200 annotated samples for learning such action. Once the new class has been learned, the transfer learning method acts as an algorithmic annotators, and labels data sequences sent from the tracker tree where incongruent actions are detected. Once the number of annotation is of at least 200 frames, the data are sent back to the tracker tree, in order to build the new action representation and integrate it in the tree. The position where the action is added to the tree depends on where in the hierarchy the incongruency has been detected.

5.3 Results

The research and development cycle within the DIRAC project, starting with the development of ideas, the description and fostering of the DIRAC paradigm of incongruency and leading to the development of detectors and models as well as the aggregation of several databases is concluded with the evaluation of these models. The databases collected by the project partners have been prepared to serve as a basis for this evaluation by annotating its content.

Evaluation of Audio-Visual Detectors

The following detectors have been evaluated against an evaluation database: the speech/non-speech detector, the acoustic localization detector, the "Walking", "Sitting" and "Picking up" detectors from the tracker tree and the Conversation detector. The evaluation database was drawn from the pooled of all project partners, sorted by content description and modality. All recordings from the evaluation database were pre-processed with the DIRAC pre-processing pipeline and processed by the detector models. For the evaluation ground truth, each relevant scene has been annotated by human inspection.

One important fact should be mentioned explicitly: no recordings from the evaluation dataset were used to train the detector models. The partners only used their own recordings which have not been part of any of the databases collected within the DIRAC project. Thus, the selected evaluation dataset are completely new to the detector models.

The different detectors evaluated for both audio and video modality showed good performance over the evaluation data set from all three recording locations. The results per location never fell below 71% for a detector per location set, and there were only subtle differences in detector performance between recordings of different locations. Over all sets and all detectors, we correctly detected a remarkable average of 75.3% of all frames.

6 Summary and Conclusions

The DIRAC project was motivated by the desire to bridge the gap between cognitive and engineering systems, and the observation that cognitive biological systems respond to unexpected events in a more robust and effective way. We first formulated a definition of rare events which goes beyond the traditional definition of event novelty. We then translated this conceptual framework to algorithms for a number of applications involving visual and auditory data. In parallel, we investigated learning mechanisms that can be used to learn the detected events. Finally, we developed a number of scenarios and collected data for which some of these algorithms could be tested with real data and in an integrated way.

References

1. Bach, J.-H., Anemüller, J.: Detecting novel objects through classifier incongruence. In: Interspeech, pp. 2206–2209 (2010)
2. Bach, J.-H., Kollmeier, B., Anemüller, J.: Modulation-based detection of speech in real background noise: Generalization to novel background classes. In: IEEE International Conference on Acoustics, Speech and Signal Processing (ICASSP), pp. 41–44 (2010)
3. De Baene, W., Premereur, E., Vogels, R.: Properties of shape tuning of macaque inferior temporal neurons examined using Rapid Serial Visual Presentation. Journal of Neurophysiology 97, 2900–2916 (2007)
4. Burget, L., Schwarz, P., Matejka, P., Hannemann, M., Rastrow, A., White, C., Khudanpur, S., Hermansky, H., Cernocky, J.: Combination of strongly and weakly constrained recognizers for reliable detection of OOVs. In: International Conference on Acoustics, Speech, and Signal Processing (ICASSP), p. 4 (2008)
5. Castellini, C., Tommasi, T., Noceti, N., Odone, F., Caputo, B.: Using object affordances to improve object recognition. IEEE Transaction on Autonomous Mental Development (2011)
6. De Baene, W., Vogels, R.: Effects of adaptation on the stimulus selectivity of macaque inferior temporal spiking activity and local field potentials. Cerebral Cortex 20(9), 2145–2165 (2010)

7. De Baene, W., Ons, B., Wagemans, J., Vogels, R.: Effects of category learning on the stimulus selectivity of macaque inferior temporal neurons. Learning and Memory 15, 717–727 (2008)
8. Deliano, Ohl: Neurodynamics of category learning: Towards understanding the creation of meaning in the brain. New Mathematics and Natural Computation (NMNC) 5, 61–81 (2009)
9. Hannemann, M., et al.: Similarity scoring for recognized repeated Out-of-Vocabulary words. In: Proc. Interspeech 2010, Makuhari, Japan (2010)
10. Hermansky, H.: Dealing With Unexpected Words in Automatic Recognition of Speech. Technical report, Idiap Research Institute (2008)
11. Herrmann, C.S., Ohl, F.W.: Cognitive adequacy in brain-like intelligence. In: Sendhoff, B., Körner, E., Sporns, O., Ritter, H., Doya, K. (eds.) Creating Brain-Like Intelligence. LNCS, vol. 5436, pp. 314–327. Springer, Heidelberg (2009)
12. Jie, L., Orabona, F., Caputo, B.: An online framework for learning novel concepts over multiple cues. In: Proceedings of Asian Conference on Computer Vision (ACCV), vol. 1, pp. 1–12 (2009)
13. Ketabdar, H., Hannemann, M., Hermansky, H.: Detection of Out-of-Vocabulary (2007)
14. Kayser, H., Ewert, S.D., Anemüller, J., Rohdenburg, T., Hohmann, V., Kollmeier, B.: Database of Multichannel In-Ear and Behind-the-Ear Head-Related and Binaural Room Impulse Responses. EURASIP Journal on Advances in Signal Processing, 1–10 (2009)
15. Words in Posterior Based ASR. In: 8th Annual Conference of the International Speech Communication Association INTERSPEECH 2007, pp. 1757–1760 (2007)
16. Kombrink, S.: OOV detection and beyond. In: DIRAC Workshop at ECML/PKDD (2010)
17. Kombrink, S., Hannemann, M., Burget, L., Heřmanský, H.: Recovery of rare words in lecture speech. In: Sojka, P., Horák, A., Kopeček, I., Pala, K. (eds.) TSD 2010. LNCS(LNAI), vol. 6231, pp. 330–337. Springer, Heidelberg (2010)
18. Kombrink, S., Burget, L., Matejka, P., Karafiat, M., Hermansky, H.: Posterior-based Out of Vocabulary Word Detection in Telephone Speech. In: ISCA, Interspeech 2009, Brighton, GB, pp. 80–83 (2009), ISSN 1990-9772
19. Mikolov, T., Karafiát, M., Burget, L., Černocký, J., Khudanpur, S.: Recurrent neural network based language model. In: Proceedings of the 11th Annual Conference of the International Speech Communication Association (INTERSPEECH 2010), pp. 1045–1048 (2010)
20. Nater, F., Grabner, H., Jaeggli, T., Gool, L.v.: Tracker trees for unusual event detection. In: ICCV 2009 Workshop on Visual Surveillance (2009)
21. Nater, F., Vangeneugden, J., Grabner, H., Gool, L.v., Vogels, R.: Discrimination of locomotion direction at different speeds: A comparison between macaque monkeys and algorithms. In: ECML Workshop on rare audio-visual cues (2010)
22. Orabona, F., Jie, L., Caputo, B.: Online-Batch Strongly Convex Multi Kernel Learning. In: Proceedings of the IEEE Computer Society Conference on Computer Vision and Pattern Recognition, CVPR 2010 (2010)
23. Orabona, F., Caputo, B., Fillbrandt, A., Ohl, F.: A Theoretical Framework for Transfer of Knowledge Across Modalities in Artificial and Biological Systems. In: IEEE 8th International Conference on Development and Learning, ICDL 2009 (2009)

24. Orabona, F., Castellini, C., Caputo, B., Luo, J., Sandini, G.: Towards Life-long Learning for Cognitive Systems: Online Independent Support Vector Machine. Pattern Recognition 43(4), 1402–1412 (2010)
25. Orabona, F., Keshet, J., Caputo, B.: Bounded kernel-based perceptrons. Journal of Machine Learning Research 10, 2643–2666 (2009)
26. Orabona, F., Keshet, J., Caputo, B.: The projectron: a bounded kernel-based perceptron. In: 25th International Conference on Machine Learning (2008)
27. Orabona, F., Castellini, C., Caputo, B., Luo, J., Sandini, G.: Indoor Place Recognition using Online Independent Support Vector Machines. In: Proceedings of the 18th British Machine Vision Conference (BMVC), pp. 1090–1099 (2007)
28. Pajdla, T., Havlena, M., Heller, J., Kayser, H., Bach, J.-H., Anemüller, J.: Incongruence Detection for Detecting, Removing, and Repairing Incorrect Functionality in Low-Level Processing (CTU-CMP-2009-19). Technical report, CTU Research Report (2009)
29. Schmidt, D., Anemüeller, J.: Acoustic Feature Selection for Speech Detection Based on Amplitude Modulation Spectrograms. In: Fortschritte der Akustik: DAGA 2007, Deutsche Gesellschaft für Akustik (DEGA), pp. 347–348 (2007)
30. Szöke, I., Fapso, M., Burget, L., Cernocky, J.: Hybrid Word-Subword Decoding for Spoken Term Detection. In: SSCS 2008 - Speech search Workshop at SIGIR, p. 4 (2008)
31. Tommasi, T., Orabona, F., Caputo, B.: Safety in numbers: learning categories from few examples with multi model knowledge transfer. In: Proceedings of the IEEE Computer Society Conference on Computer Vision and Pattern Recognition, CVPR 2010 (2010)
32. Tommasi, T., Caputo, B.: The more you know, the less you learn: from knowledge transfer to one-shot learning of object categories. In: British Machine Vision Conference, BMVC 2009 (2009)
33. Vangeneugden, J., De Mazière, P., Van Hulle, M., Jaeggli, T., Van Gool, L., Vogels, R.: Distinct Mechanisms for Coding of Visual Actions in Macaque Temporal Cortex. Journal of Neuroscience 31(2), 385–401 (2011)
34. Vangeneugden, J., Vancleef, K., Jaeggli, T., Van Gool, L., Vogels, R.: Discrimination of locomotion direction in impoverished displays of walkers by macaque monkeys. Journal of Vision 10(4), 22.1–22.19 (2010)
35. Vangeneugden, J., Pollick, F., Vogels, R.: Functional differentiation of macaque visual temporal cortical neurons using a parametric action space. Cerebral Cortex 19(3), 593–611 (2009)
36. Verhoef, B.E., Kayaert, G., Franko, E., Vangeneugden, J., Vogels, R.: Stimulus similarity-contingent neural adaptation can be time and cortical area dependent. Journal of Neuroscience 28, 10631–10640 (2008)
37. White, C., Zweig, G., Burget, L., Schwarz, P., Hermansky, H.: Confidence Estimation, Oov Detection And Language Id Using Phone-To-Word Transduction And Phone-Level Alignments. In: IEEE Int. Conf. on Acoustics, Speech, and Signal Processing, pp. 4085–4088 (2008)
38. Witte, H., Charpentier, M., Mueller, M., Voigt, T., Deliano, M., Garke, B., Veit, P., Hempel, T., Diez, A., Reiher, A., Ohl, F., Dadgar, A., Christen, J., Krost, A.: Neuronal cells on GaN-based materials. Deutsche Physikalische Gesellschaft, Spring Meeting of the Deutsche Physikalische Gesellschaft, Berlin (2008)

Part II
The Detection of Incongruent Events, Project Survey and Algorithms

Audio Classification and Localization for Incongruent Event Detection

Jörg-Hendrik Bach, Hendrik Kayser, and Jörn Anemüller

Abstract. A method is presented that detects unexpected acoustic events, i.e., occurrence of acoustic objects that do not belong to any of the learned classes but nevertheless appear to constitute meaningful acoustic events. Building on the framework [Weinshall et al.], general and specific acoustic classifiers are implemented and combined for detection of events in which they respond in an incongruous way, indicating an unexpected event. Subsequent identification of events is performed by estimation of source direction, for which a novel classification-based approach is outlined. Performance, evaluated in dependence of signal-to-noise ratio (SNR) and type of unexpected event, indicates decent performance at SNRs better than 5 dB.

1 Introduction

Human detection of acoustic objects and events is remarkably efficient even when sounds of interest occur within an uninteresting floor of background activity. Further, we are also very good at detecting acoustic sounds even when they belong to a class we have not encountered before. In such a situation, two questions need to be answered simultaneously: Does the present acoustic scene contain a significant event at all? And if so: Does the event belong to any of the event classes know from previous experience? If the latter is not the case, the event is termed "unexpected", and commonly subsequent actions for its further identification and description ensue, such as localizing the direction in which the event occurred, increasing attention towards this direction, etc.

Jörg-Hendrik Bach · Hendrik Kayser · Jörn Anemüller
Carl von Ossietzky Universität Oldenburg,
Medical Physics,
26111 Oldenburg, Germany
e-mail: {j.bach,hendrik.kayser}@uni-oldenburg.de,
     joern.anemueller@uni-oldenburg.de
     http://medi.uni-oldenburg.de

D. Weinshall, J. Anemüller, and L. van Gool (Eds.): DIRAC, SCI 384, pp. 39–46.
springerlink.com   © Springer-Verlag Berlin Heidelberg 2012

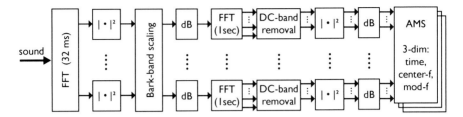

Fig. 1 AMS feature extraction.

The present contribution outlines a system for performing the unexpected event detection task in the auditory modality. It builds on the general framework based on levels of ("general" and "specific") classifiers and detection of incongruences thereof, put forth in [7]. Implementation of such a system necessitates the specification of several processing modules, each of which is presented in the present work: Signal pre-processing is outlined in sections 2.1 and 2.2, yielding salient feature sets for classification and localization of acoustic events. The incongruous event detection framework is summarized in section 2.3 and the resulting audio classifier architecture presented. Section 3 highlights results of detection and localization of events.

2 Methods

2.1 Object Detection

Feature Extraction

The features used in this task are Amplitude Modulation Spectrograms (AMS), which are motivated by the importance of temporal modulations for acoustic object recognition. This importance has been confirmed in numerous psychophysical, physiological and applied studies (e.g. [6]). AMS represent a decomposition of the signal along the dimensions of acoustic frequency, modulation frequency and time, and are computed by a (modulation) spectral decomposition of sub-band spectral power time-courses in overlapping temporal windows. The processing stages of the AMS computation are as follows (see Fig. 1). The signal decomposition with respect to acoustic frequency is computed by a short-term fast Fourier transformation (FFT with 32 ms Hann window, 4 ms shift, FFT length 256 samples, sampling rate 8 kHz.), followed by squared magnitude computation, summation into rectangular, non-overlapping Bark bands and logarithmic amplitude compression. Within each spectral band, the modulation spectrum is obtained by applying another FFT (1000 ms Hann window, 500 ms shift, FFT length 250 samples) to the temporal trajectories of subband log-energy. Finally, an envelope extraction and a further logarithmic compression are applied. The resolution of 17 Bark acoustic frequency bands and 29 modulation frequency bands (2 Hz to 30 Hz) results in a signal representation with

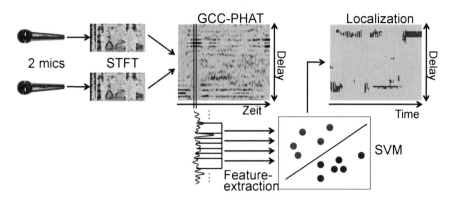

Fig. 2 Schematic of the classification approach for sound source localization.

493 feature-values per time step, at a temporal resolution of 1000 ms. Robustness of the AMS features with respect to constant and slowly-varying time-domain convolutive factors (channel noise) is enhanced by construction of the feature extraction scheme: The first Fourier transform step approximately converts the convolution to a multiplicative term in each (acoustic) frequency band. The subsequent logscale compression converts it into an additive term that is captured in the DC (0 Hz) modulation spectral band after the second FFT (depending on parameter choice, additional low modulation-frequency may be influenced by smearing of DC components). Deliberately discarding the affected modulation bands therefore results in AMS features that are approximately invariant to a time-domain signal convolution with short impulse responses such as microphone transfer functions and early reverberation effects.

Classification

The object detection task is mapped onto a discrimination task as follows: a number of known objects is classified against one another using a 1-vs-all discriminative approach. Support Vector Machines [3] are employed as a back-end. For each object, a binary classifier is trained to discriminate between "target object in background" and "different object in background".

2.2 Localization

The classification approach for sound source localization (sketched in Fig. 2) employs features extracted from the generalized cross correlation (GCC) function [5] between two audio input signals. The GCC encodes information about the relative temporal delay with which sound waves originating from a sound source at a certain position impinge on the different microphones. A sound source is localized by estimating the corresponding azimuthal angle of the direction of arrival (DOA) relative to the sensor plane defined by the two front-microphones of the AWEAR platform which is an mobile audio-visual recording device developed within the DIRAC project. Support-vector-machines (SVM) [3] are used to classifiy the presence or absence of a source at each angle. This approach enables the simultaneous localization of more than one sound source in each time-frame.

The GCC is an extension of the cross power spectral density function, which is given by the Fourier transform of the cross correlation. Given two signals $x_1(n)$ and $x_2(n)$, it is defined as:

$$G(n) = \frac{1}{2\pi} \int_{-\infty}^{\infty} H_1(\omega)H_2^*(\omega) \cdot X_1(\omega)X_2^*(\omega)e^{j\omega n}\, d\omega, \tag{1}$$

where $X_1(\omega)$ and $X_2(\omega)$ are the Fourier transforms of the respective signals and the term $H_1(\omega)H_2^*(\omega)$ denotes a general frequency weighting.

In the present work PHAse Transform (PHAT) weighting [5] has been used, which normalizes the amplitudes of the input signals to unity in each frequency band, $H_1(\omega)H_2^*(\omega) = \frac{1}{|X_1(\omega)X_2^*(\omega)|}$:

$$G_{PHAT}(n) = \frac{1}{2\pi} \int_{-\infty}^{\infty} \frac{X_1(\omega)X_2^*(\omega)}{|X_1(\omega)X_2^*(\omega)|} e^{j\omega n}\, d\omega, \tag{2}$$

such that only the phase difference between the input signals is preserved.

The inter-microphone distance in the AWEAR setup (45 cm) corresponds to a maximum delay of 1.32 ms ($\widehat{=} \pm 90°$) in each direction. The window length and the length of the Fourier transform used to compute the GCC in the spectral domain are chosen such that a 257-dimensional GCC-PHAT vector is produced for every time frame of 5.33 ms length.

From the GCC data, angular features are extracted for classification in the following way: A rectangular window is slid over the GCC-vector, subdividing the data of a single time frame into 61 feature-vectors. This covers the field of view homogeneously in terms of time delay with a resolution of $\frac{1}{3}$ ms. The mapping from the time delay τ to the angle of incidence θ is non-linear: $\theta = \arcsin(\tau \cdot \frac{c}{d})$, where c denotes the speed of sound and d the distance between the sensors. This results in a non-homogenous angular resolution in the DOA-angle-space, with higher resolution near the centre and lower resolution towards the edges of the field of view.

2.3   Incongruous Event Detection

Framework

We use the concept of incongruence proposed in [7] to detect unexpected events in acoustic scenes. In a nutshell, an incongruence is defined as a disagreement between models of different levels of abstraction. An event is considered incongruous if the

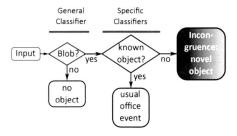

Fig. 3 Flow chart of the novelty detection based on incongruence between models for general and specific acoustic objects.

classifier for the more abstract (general) concept is confident about its classification, whereas the less abstract (specific) classifier is not. This allows the detection of events that are unknown to the combination of classifiers while not necessarily novel to each individual classifier.

Implementation for Audio Signals

In previous work, we focussed on detecting known acoustic objects in unknown background sounds [2]. Here we aim at detecting unknown ("incongruent") objects in known background sounds using general and specific object classifiers.

The general object detector is based on a RASTA-PLP feature extraction [4]. The PLPs were extracted using 17 Bark bands, 25 ms windows with 10 ms shift and the original log-RASTA filter. For each time frame, a positive vote for the presence of an object ("blob") is collected if the sum over all coefficients is above a certain threshold θ. If at least 5% of the votes within a 1 s window are positive, that window is classified as containing an acoustic blob. The detector is evaluated by varying the threshold θ above which an input frame is classified as 'object present'. The test was performed using the background sound for the 'no object' class and all known office objects as 'object present'. Hit rates and false alarms derived from this test are summarized in receiver operating characteristics (ROC). False positives refer to background wrongly classified as blob.

The specific detectors use amplitude modulation spectrograms (AMS, see Section 2.1) as input features. Using these features, support vector machines (SVM, [3]) are trained as models for the specific acoustic objects. The training is performed using a 1-vs-all approach. The performance of the specific models has been evaluated as follows: each of the four office objects is defined as *new* once and left out of the training set. This new object is detected by running the data through the classifiers for all *known* objects. Because of this procedure, separate models are trained for each defined train set. For example, when using speech as novel event, the door detector is trained as "door vs. {keyboard,telephone}".

3 Results

3.1 Data

The data for all classes (including the office background noise) except speech has been recorded in a typical office at the University of Oldenburg. Separate recording sessions of approximately 15 min each have been used for train and test data. The office background noise is dominated by an air conditioning ventilation system and comparatively stationary. Speech data is provided by the TIMIT database. The sound objects have been mixed into the background at SNRs from $+20$ dB to -20 dB using a long-term (over the whole signal) and broad band SNR computation.

To produce a suitably large set of training data for the localization model, sound sources were simulated by using mono-channel speech recordings from the TIMIT speech corpus and generating a second channel as a delayed version of the same data, thus introducing directional information into the data. The simulated DOA angle ranged from $-80°$ (left) to $+80°$ (right) in steps of $10°$, for each direction 10 seconds of speech were used. A spherically symmetric diffuse pink noise field was generated and superposed on the speech signals with varying long-term SNR ranging from -20 dB to +20 dB in steps of 5 dB.

3.2 Detection of Incongruous Events

The general model (acoustic blob detector) has been tested on the office data test set. All four objects as well as the background sounds have been used, and the full test set has been balanced to ensure equal amounts of data for 'object present' and 'background only'. The results are summarized in Table 1, see [1] for a more detailed exposition.

The results of the specific classifiers for detection of speech as incongruent event are shown in Fig. 4, left panel. These are combined with the result of the general detector to an overall detection score for new objects as follows: the ROC performance of the specific models is scaled by the hit rate of the general detector, reflecting the fact that in the full system, the specific models are fed only those data points specified as blobs by the general detector (see Fig. 3). This is equivalent to rescaling the ROC axes from $[0, 1]$ to $[0, \text{blob hit rate}]$. The accuracy at the EER as a function of the SNR is given in Fig. 4, right panel.

The detection of incongruous events works well ($> 75\%$ detection rate on average) down to 0 dB and approximates chance level at -10 dB. At SNRs below -10 dB, the hit rate of the blob detection is too low, preventing the rescaled ROC of the overall performance to cross the EER line, therefore no EER values can be given.

In summary, the results indicate that the framework used is suitable for novelty detection based on incongruence of different models.

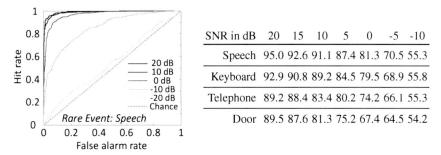

SNR in dB	20	15	10	5	0	-5	-10
Speech	95.0	92.6	91.1	87.4	81.3	70.5	55.3
Keyboard	92.9	90.8	89.2	84.5	79.5	68.9	55.8
Telephone	89.2	88.4	83.4	80.2	74.2	66.1	55.3
Door	89.5	87.6	81.3	75.2	67.4	64.5	54.2

Fig. 4 Left panel: ROC performance of the specific models at different SNRs, testing with speech as the incongruent object. Right panel: Accuracy of the detection of incongruous events taken at the EER point. Below −10 dB, the EER could not be determined.

Table 1 Accuracy of the blob detector taken at the EER point (equal false alarm and miss rates).

SNR in dB	20	15	10	5	0	-5	-10	-15	-20
Accuracy in %	97.4	96.5	94.5	90.6	86.7	80.3	73.4	64.5	56.9

3.3 Localization of Events

The evaluation showed a high robustness of the approach. the performance is stable under mismatch of training and test SNRs and independent of the direction from which a sound source impinges. Results for a single sound source are shown in Fig. 5(a).

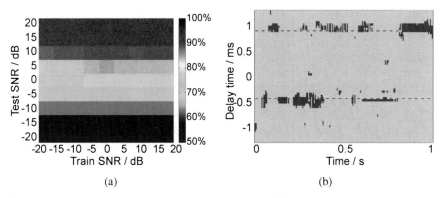

(a)   (b)

Fig. 5 (a) performance of the localization algorithm for different and mismatched train and test snrs. the percentage of correct decisions (source / no source) is shown for a single sound source. (b) results of the classification for two speech sources at −20° and +45° in a isotropic noise field at an snr of 5 db. the dotted lines in the right plot denote the theoretical positions of the sources.

A classification example for two speech sources with interference by a noise field at an SNR of 5 db is shown in Fig. 5(b). The ground truth positions of the two speech sources at $-20°$ and $+45°$ are indicated by the dotted lines in the right panel. To suppress the "salt and pepper" noise contained in the results of the classification, median filtering over 3 adjacent angles and 7 time frames was applied.

4   Conclusion

We have presented a system for detection and identification of unexpected events in the auditory modality, building on the theory of incongruous event detection [7] and implementing it using classifiers optimized for acoustic signals. Classification built on the hierarchical classification scheme with general and specific classifiers incorporating previously developed acoustic classification methods. Identification of events subsequent to their detection was done using a novel localization algorithm that performs classification-based estimation of source direction.

Results indicate that the system can successfully detect and localize incongruous events when the SNR is better than 5 dB, in which case equal-error-rate detection performance exceeds between 74% and 81%, depending on the type of incongruous event. At beneficial SNR (20 dB), performance exceeds 89% to 95%. The data also demonstrate that the given numbers reflect performance of the combined scheme, i.e., general (blob) classifier, specific (object) classifiers and detection of incongruence between them. Hence, further work will address all three parts to further optimize performance.

References

1. Bach, J.H., Anemüller, J.: Detecting novel objects through classifier incongruence. In: Proc. Interspeech, Makuhari, Japan, pp. 2206–2209 (2010)
2. Bach, J.H., Kollmeier, B., Anemüller, J.: Modulation-based detection of speech in real background noise: Generalization to novel background classes. In: Proceedings of the International Conference on Acoustics, Speech, and Signal Processing (ICASSP), Dallas, Texas, pp. 41–45 (2010)
3. Chang, C.C., Lin, C.J.: LIBSVM: a library for support vector machines (2001), software available, http://www.csie.ntu.edu.tw/~cjlin/libsvm
4. Hermansky, H., Morgan, N.: Rasta processing of speech. IEEE Transactions on Speech and Audio Processing 2(4), 578–589 (1994)
5. Knapp, C., Carter, G.: The generalized correlation method for estimation of time delay. IEEE Transactions on Acoustics, Speech and Signal Processing 24(4), 320–327 (1976)
6. Tchorz, J., Kollmeier, B.: SNR estimation based on amplitude modulation analysis with applications to noise suppression. IEEE Transactions on Speech and Audio Processing 11(3), 184–192 (2003)
7. Weinshall, D., Hermansky, H., Zweig, A., Luo, J., Jimison, H., Ohl, F., Pavel, M.: Beyond novelty detection: Incongruent events, when general and specific classifiers disagree. In: Koller, D., Schuurmans, D., Bengio, Y., Bottou, L. (eds.) Advances in Neural Information Processing Systems, vol. 21, pp. 1745–1752 (2009)

Identification of Novel Classes in Object Class Recognition

Alon Zweig, Dagan Eshar, and Daphna Weinshall

Abstract. For novel class identification we propose to rely on the natural hierarchy of object classes, using a new approach to detect incongruent events. Here detection is based on the discrepancy between the responses of two different classifiers trained at different levels of generality: novelty is detected when the general level classifier accepts, and the specific level classifier rejects. Thus our approach is arguably more robust than traditional approaches to novelty detection, and more amendable to effective information transfer between known and new classes. We present an algorithmic implementation of this approach, show experimental results of its performance, analyze the effect of the underlying hierarchy on the task and show the benefit of using discriminative information for the training of the specific level classifier.

1 Introduction

A number of different methods have been developed to detect and recognize object classes, showing good results when trained on a wide range of publicly available datasets (see e.g. [1, 3]). These algorithms are trained to recognize images of objects from a known class. Ideally, when an object from a new class appears, all existing models should reject it; this is the only indication that a new object class has been seen. However, similar indication will be obtained when no object exists in the image. And what with outliers and noisy images, low recognition likelihood by all existing models may be obtained even when a known object is seen in the image. This is one of the fundamental difficulties with the prevailing novelty detection paradigm - negative evidence alone gives rather non-specific evidence, and may be the result of many unrelated reasons.

In this paper we are interested in novel object class identification - how to detect an image from a novel class of objects? Unlike the traditional approach to novelty detection (see [4, 5] for recent reviews), we would like to utilize the natural

Alon Zweig · Dagan Eshar · Daphna Weinshall
School Computer Science & Engineering, Hebrew University of Jerusalem, Jerusalem, Israel

D. Weinshall, J. Anemüller, and L. van Gool (Eds.): DIRAC, SCI 384, pp. 47–55.
springerlink.com

hierarchy of objects, and develop a more selective constructive approach to novel class identification. Our proposed algorithm uses a recently proposed approach to novelty detection, based on the detection of incongruencies [6]. The basic observation is that, while a new object should be correctly rejected by all existing models, it can be recognized at some more abstract level of description. Relying on positive identification at more abstract levels of representation allows for subsequent modeling using related classes, as was done e.g. in [7], where the representation of a new class was built from a learnt combination of classifiers at different levels of generality.

Specifically, in Section 2 we describe our approach to novel class identification. We consider a multi-class scenario, where several classes are already known; a sample from an unknown class is identified based on the discrepancy between two classifiers, where one accepts the new sample and the second rejects it. The two classifiers are hierarchically related: the accepting classifier fits a general object class description, and the rejecting classifier fits the more specific level describing the known object classes. We use this property, that detection is based on rejection by only a small number of related classes, to develop a discriminative approach to the problem - the specific classifiers are trained to distinguish between the small group of related sub-classes in particular.

Our approach is general in the sense that it does not depend on the specifics of the underlying object class recognition algorithm. To investigate this point, we used in our implementation (and experiments) two very different publicly available object recognition methods [1, 3]. Due to space limitations we present results only for [1], applying our approach when using the model of [3] showed similar results.

We tested our algorithms experimentally (see Section 3) on two sets of objects: a facial data set where the problem is reduced to face verification, and the set of motorbikes from the Caltech256 bench-mark dataset. We found that discriminative methods, which capture distinctions between the related known sub-classes, perform significantly better than generative methods. We also demonstrate in our experiments the importance of modeling the hierarchical relations as tightly as possible.

2   Our Approach and Algorithm

In order to identify novel classes, we detect discrepancies between two levels of classifiers that are hierarchically related. The first level consists of a single 'general category' classifier, which is trained to recognize objects from any of the known sub-classes, see Section 2.1. The second level is based on a set of classifiers, each trained to make more specific distinctions and classify objects from a small group of related known sub-classes. Using these specific classifiers, we build a single classifier which recognizes a new sample as either belonging to one of the set of known sub-classes or not, see Section 2.2.

We look for a discrepancy between the inference made by the two final classifiers, from the two different levels, where the general classifier accepts and the specific classifier rejects a new sample. This indicates that the new sample belongs to the general category but not to any specific sub-class; in other words, it is a novel class.

This algorithm is described a bit more formally in Algorithm 1, with further details given in the following subsections.

Algorithm 1. Unknown Class Identification

Input :

x test image
C_G general level classifier
C_j specific level classifiers, $j = 1..|\#of\text{known sub-classes}|$
$V_{C_i}^c$ average confidence of train or validation examples classified correctly as C_i
$V_{C_i}^w$ average confidence of train or validation examples classified wrongly as C_i (zero if there are none).

1. Classify **x** using C_G
2. if accept
   Classify **x** using all C_j classifiers and obtain a set of confidence values $V_{C_j}(\mathbf{x})$
   Let $i = \arg\max\limits_{j} V_{C_j}(\mathbf{x})$
   Define $S(\mathbf{x}) = (V_{C_i}(\mathbf{x}) - V_{C_i}^w)/(V_{C_i}^c - V_{C_i}^w)$

   a. if $S(\mathbf{x}) > 0.5$
      label **x** as belonging to a known class
   b. else label **x** as belonging to a novel (unknown) class

3. else label **x** as a background image

2.1 General Category Level Classifier

In order to learn the general category level classifier, we consider a small set of related known classes as being instances of a single (higher level, or more abstract) class. In accordance, all the examples from the known classes are regarded as the positive set of training examples for the more abstract class. For the negative set of examples we use either clutter or different unrelated objects (none of which is from the known siblings). As we shall see in Section 3, this general classifier demonstrates high acceptance rates when tested on the novel sub-classes.

2.2 Specific Category Level Classifier

At the specific level classifier the problem is essentially reduced to the standard novelty detection task of deciding whether a new sample belongs to any of the known classes or to an unknown class. However, the situation is somewhat unique and we take advantage of this: while there are multiple known classes, their number is bounded by the degree of the hierarchical tree (they must all be sub-classes of a single abstract object). This suggests that a discriminative approach could be rather effective.

The training procedure of the specific level classifier is summarized in Algorithm 2 with details provided in subsequent subsections, while the classification process is described in step 2 of Algorithm 1 above.

Algorithm 2. Train Known Vs. Unknown Specific Class Classier

1. For each specific class, build a discriminative classifier with: *positive examples:* all images from the specific class. *negative examples:* images from all sibling classes.
2. Compute the Normalized Confidence function.
3. Choose a classification threshold for novel classes.

Step 1: Discriminative Multi-Class Classification. To solve the multi-class classification problem in Step 1 of the algorithm, we train a discriminative object class classifier for each of the known specific classes and classify a new sample according to the most likely classification (max decision). We incorporate the discriminative information in the training phase of each of the single known classes-specific classifiers. Specifically, each single class classifier is trained using all images from its siblings (other known classes under the same general level class) as negative examples. Thus the specific level object model learnt for each known class is optimized to separate the class from its siblings.

Step 2: Normalized Confidence Score. For a new sample $\mathbf{x}$ which is classified as C_i, we obtain an estimate for classification confidence $V_{C_i}(\mathbf{x})$, the output of the learnt classifier. After the max operation in the Step 1, this value reflects the classification confidence of the multi-class classifier. Given this estimate, we would like to derive a more accurate measure of confidence as to whether or not the classified sample belongs to the group of known sub-classes.

To do this, we define a normalized score function, which normalizes the confidence estimate $V_{C_i}(\mathbf{x})$ relative to the confidence estimates of correct classifications and wrong classifications for the specific-class classifier, as measured during training or validation. Specifically, let $V_{C_i}^c$ denote the average confidence of train or validation examples classified correctly as C_i, and let $V_{C_i}^w$ denote the average confidence of train or validation examples from all other sub-classes classified wrongly as belonging to class C_i. The normalized score $S(\mathbf{x})$ of $\mathbf{x}$ is calculated as follows:

$$S(\mathbf{x}) = \frac{(V_{C_i}(\mathbf{x}) - V_{C_i}^w)}{(V_{C_i}^c - V_{C_i}^w)} \tag{1}$$

If the classes can be well separated during training, that is $V_{C_i}^c >> V_{C_i}^w$ and both groups have low variance, the normalized score provides a reliable confidence measure for the multi-class classification.

Step 3: Choosing a threshold. Unlike the typical discriminative learning scenario, where positive and negative examples are given during training, in the case of novelty detection no actual negative examples are known during training. In particular, when the specific-object classifiers are trained, they are given no example of the new sub-class whose detection is the goal of the whole process. Thus it becomes advantageous to set rather conservative limits on the learnt classifiers, more so than indicated by the train set. In other words, in order to classify a new sample as known (a positive example for the final classifier), we will look for higher confidence in the classification. This is done by setting the threshold of the normalized confidence measure to reject more than originally intended during training. Since the normalized confidence measure lies in the range $[0..1]$, we set the threshold in our experiments to 0.5.

3 Experiments

3.1 Datasets and Method

We used two different hierarchies in our experiments. In the first hierarchy, the general parent category level is the 'Motorbikes'. 22 object classes, taken from [2], were added, in order to serve together with the original data set as the pool of object classes used both for the unseen-objects results described in section 3.3 and for the random grouping described in Section 3.5. In the second hierarchy, the general parent category level is the 'Face' level, while the more specific offspring levels are faces of six different individuals.

All experiments were repeated at least 25 times with different random sampling of test and train examples. We used 39 images for the training of each specific level class in the 'Motorbikes' hierarchy, and 15 images in the 'Faces' hierarchy. For each dataset with n classes, n conditions are simulated, leaving each of the classes out as the unknown (novel) class.

3.2 Basic Results

Figure 1 shows classification results for the discriminative approach described in Section 2. These results show the classification rates for the different types of test samples: *Known* - samples from all known classes during the training phase; *Unknown* - samples from the unknown (novel) class which belongs to the same General level as the Known classes but has been left out during training; *Background* - samples not belonging to the general level which were used as negative examples during the General level classifier training phase; and *Unseen* - samples of objects from classes not seen during the training phase, neither as positive nor as negative examples. The three possible types of classification are: *Known* - samples classified as belonging to one of the known classes; *Unknown* - samples classified as belonging to the unknown class; and *Background* - samples rejected by the General level classifier.

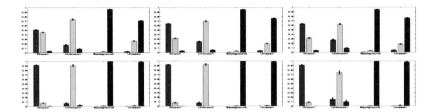

Fig. 1 Classification ratios for 4 groups of samples: Known Classes, Unknown Class, Background and sample of unseen classes. Ratios corresponding to the three possible classification rates are shown: left bar (blue) shows the known classification rate, middle bar (green) shows the unknown classification rate, and right bar (red) shows the background classification rate (rejection by the general level classifier). The top row plots correspond to the Motorbikes general level class, where the Cross (left), Sport (middle) and Road Motorbikes (right) classes are each left out as the unknown class. The bottom row plots are representative plots of the Faces general level class, where KA (left), KL (middle) and KR (right) are each left out as the unknown class.

The results in Fig. 1 show the desired effects: each set of samples - Known, Unknown and Background, has the highest rate of correct classification as its own category. As desired, we also see similar recognition rates (or high acceptance rates) of the Known and Unknown classes by the general level classifier, indicating that both are regarded as similarly belonging to the same general level. Finally, samples from the Unseen set are rejected correctly by the general level classifier.

3.3 Discriminative Specific Classifiers Improve Performance

We checked the importance of using a discriminative approach by comparing our approach for building discriminative specific-level classifiers to non-discriminative approaches. In all variants the general level classifier remains the same.

We varied the amount of discriminative information used when building the specific level classifiers, by choosing different sets of examples as the negative training set: 1) *1vsSiblings - Exploiting knowledge of sibling relations*, the most discriminative variant, where all train samples of the known sibling classes are used as the negative set when training each specific known class classifier. 2) *1vsBck - No knowledge of siblings relations*, a less discriminative variant, where the negative set of examples is similar to the one used when training the general level classifier.

Applying these different variants when training models of [1] results in entirely different object models for which known vs. unknown classification results depict different ROC curves, as shown in Fig. 2. Specifically, when comparing the *1vsSiblings* to *1vsBck* curves in Fig. 2, we see that for all different choices of classes left out as the unknown - the corresponding ROC curve of the *1vsSiblings* method shows much better discrimination of the known and unknown classes. This demonstrates that discriminative training with the sibling classes as negative samples significantly enhances performance.

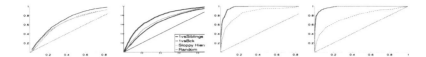

Fig. 2 ROC curves showing True-Unknown classification rate on the Y-axis vs. False-Unknown Classification rate on the X-axis. We only plot examples accepted by the General level classifier. *1vsSiblings* denotes the most discriminative training protocol, where specific class object models are learnt using the known siblings as the negative set. *1vsBck* denotes the less discriminative training protocol where the set of negative samples is the same as in the training of the General level classifier. Sloopy-Hierarchy denotes the case where the hierarchy was built using the procedure described in Section 3.5. Each plot shows results for a different class left out as the unknown, from left to right and top to bottom respectively: 'Cross-Motorbikes' , 'Sport-Motorbikes', 'KA' Face and 'KL' Face. We only show two representative cases for each dataset, as the remaining cases look very similar.

3.4 Novel Class Detector Is Specific

To test the validity of our novel class detection algorithm, we need to verify that it does not mistakenly detect low quality images, or totally unrelated novel classes, as novel sub-classes. Thus we looked at two types of miss-classifications:

First, we tested our algorithm on samples of objects from classes which are not related to the general class and had not been shown during the training phase. These samples are denoted *unseen*, for which any classification other than rejection by the general level classifier is false. We expect most of these objects to be rejected, and expect the rate of false classification of unrelated classes as unknown classes to be similar to the rate of false classification as known classes. As Fig. 1 shows, by far most unseen samples are correctly rejected by the general level classifier. For the Faces data set we see that in cases of miss-classification there is no tendency to prefer the unknown class, but this is not the case with the Motorbikes data set; thus our expectation is only partially realized. Still, most of the false classification can be explained by errors inherited from the embedded classifiers.

Second, to test the recognition of low quality images, we took images of objects from known classes and added increasing amounts of Gaussian white noise to the images. As can be seen in Fig. 1, almost all the background images are rejected correctly by the general level classifier, and the addition of noise maintains this level of rejection. On the other hand the fraction of known objects classified correctly decreases as we increase the noise.

In Fig. 3 we examine the pattern of change in the misclassification of samples from the known class with increasing levels of noise - whether a larger fraction is misclassified as an unknown object class or as background. Specifically, we show the ratio (FU-FB)/(FU+FB), where FU denotes false classification as unknown class, and FB denotes false classification as background. The higher this ratio is, the higher the ratio of unknown class misclassifications to background misclassifications is. An increase in the false identification of low quality noisy images as the unknown class should correspond to an increase in this expression as the noise increases. In fact, in Fig. 3 we see the opposite - this expressions decreases with noise. Thus, at least as it concerns low quality images due to Gaussian noise, our model does not identify these images as coming from novel classes.

Fig. 3 This figure shows the effect of noise on the rate of false classification of samples from known classes. Each bar shows the average over all experiments of (FU-FB)/(FU+FB). Results are shown for both the Motorbikes and Faces datasets, and each group of bars shows results for a different class left out as the unknown. In each group of bars, the bars correspond to increasing levels of noise, from the leftmost bar with no noise to the fifth rightmost bar with the most noise.

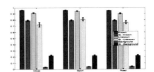

Fig. 4 General level classifier acceptance rates, with Strict and sloppy hierarchies. Six bars show, from left to right respectively: Strict hierarchy known classes ('Known'), Sloppy hierarchy ('SL-Known'), Strict hierarchy unknown class ('Unknown'), Sloppy hierarchy ('SL-Unknown'), Strict hierarchy background ('Background'). Sloppy hierarchy ('SL-Background'). Results are shown for the cases where the Cross, Sport or Road-Motorbikes are left as the unknown class, from left to right respectively

3.5   Sloppy Hierarchical Relation

In order to explore the significance of hierarchy in our proposed scheme, we followed the procedure as described in section 2 using a "sloppy" hierarchy, thus comparing results with "strict" hierarchy to results with "sloppier" hierarchy. We only changed one thing - instead of using a group of strict hierarchically related subclasses, we collected a random group of sub-classes; all other steps remained unchanged. The acceptance rate by the general level classifier using the strictly built hierarchy vs. sloppy hierarchy is shown in Fig. 4. Results are shown for objects belonging to the known classes, unknown-class and background images. Correct unknown classification vs. false unknown classification of samples that were accepted by the general level classifier are shown in Fig. 2 for both the strict and sloppy hierarchy.

As can be seen in Fig. 4, the general level classifier which is learnt for the sloppy hierarchy is less strict in the sense that more background images are falsely accepted and more known and unknown images are falsely rejected by the general level classifier. We also see from Fig. 2 that the distinction between known classes and an unknown class is significantly improved with the strict hierarchy as compared to the sloppy hierarchy. Combining both the general level classifier and the specific level classifier, clearly Algorithm 1 for the identification of unknown classes performs better when given access to a strict hierarchy, as compared to some sloppier hierarchy.

4 Summary

We address the problem of novel object class recognition - how to know when confronted with the image of an object from a class we have never seen before. We exploit existing hierarchical relations among the known classes, and propose a hierarchical discriminative algorithm which detects novelty based on the disagreement between two classifier: some general level classifier accepts the new image, while specific classifiers reject it. We analyze the properties of the algorithm, showing the importance of modeling the hierarchical relations as strictly as possible, and the importance of using a discriminative approach when training the specific level classifier.

References

1. Bar-Hillel, A., Hertz, T., Weinshall, D.: Efficient learning of relational object class models. In: ICCV (2005)
2. Griffin, G., Holub, A., Perona, P.: Caltech-256 object category dataset. Technical Report UCB/CSD-04-1366, California Institute of Technology (2007)
3. Leibe, B., Leonardis, A., Schiele, B.: Robust object detection with interleaved categorization and segmentation. IJCV 77(1), 259–289 (2008)
4. Markou, M., Singh, S.: Novelty detection: a review-part 1: statistical approaches. Signal Processing 83(12), 2481–2497 (2003)
5. Markou, M., Singh, S.: Novelty detection: a review-part 2: neural network based approaches. Signal Processing 83(12), 2499–2521 (2003)
6. Weinshall, D., Hermansky, H., Zweig, A., Luo, J., Jimison, H., Frank, O., Povel, M.: Beyond Novelty Detection: Incongruent Events, when General and Specific Classifiers Disagree. In: NIPS (2008)
7. Zweig, A., Weinshall, D.: Exploiting Object Hierarchy: Combining Models from Different Category Levels. In: ICCV (2007)

Out-of-Vocabulary Word Detection and Beyond

Stefan Kombrink, Mirko Hannemann, and Lukáš Burget

Abstract. In this work, we summarize our experiences in detection of unexpected words in automatic speech recognition (ASR). Two approaches based upon a paradigm of incongruence detection between generic and specific recognition systems are introduced. By arguing, that detection of incongruence is a necessity, but does not suffice when having in mind possible follow-up actions, we motivate the preference of one approach over the other. Nevertheless, we show, that a fusion outperforms both single systems. Finally, we propose possible actions after the detection of unexpected words, and conclude with general remarks about what we found to be important when dealing with unexpected words.

1 Unexpected Events in Speech Recognition

Events in speech can be arbitrary sounds. One possible challenge is to decide whether a particular sound is actually speech or noise and is called speech/non-speech or voice activity detection (VAD). Another challenge is to find the most likely sequence of words given a recording of the speech and a speech/non-speech segmentation. This is commonly known as automatic speech recognition (ASR) where words are constructed as a sequence of speech sounds (usually phonemes).

Although the set of speech sounds is considered to be limited, the set of words is not (in general, there is no known limit for the length of a word). Language models are commonly used in ASR to model prior knowledge about the contextual relationship of words within language. This prior probability distribution over words is conditioned on a history of preceding words and highly skewed. Usually, this distribution is discrete, i.e. only a limited set of

Stefan Kombrink · Mirko Hannemann · Lukáš Burget
Brno University of Technology, CZ
e-mail: {kombrink,ihannema,burget}@fit.vutbr.cz

D. Weinshall, J. Anemüller, and L. van Gool (Eds.): DIRAC, SCI 384, pp. 57–65.

most frequent words is known to the system. Unknown words constitute an unexpected event, and since most words occur rarely, enlarging the vocabulary does not alleviate this effect. In fact, the recognizer will replace each of these so-called out-of-vocabulary (OOV) words by a sequence of similar sounding in-vocabulary (IV) words, thus increasing the number of word errors and leading to loss of information.

Here, we investigate two different approaches to detect OOV words in speech sharing a similar strategy: finding incongruences between the output of a generic (unconstrained) and a specific (constrained by prior knowledge) system. We combine both approaches, show results of the fusion, and interpret those. The final part is dedicated to possible follow-up actions after OOVs have been detected.

2   Neural Network Based OOV Word Detection System

If the outcome of an unbiased observation contradicts the expectations raised by higher level knowledge, we refer to this as an incongruent event. The incongruence can be detected by comparing the output of a generic and a specific recognizer. In our case, the specific recognizer uses prior knowledge in form of a language model, vocabulary and pronunciation dictionary, and searches for the best sequence of words with the highest overall likelihood. The generic recognizer uses only a limited temporal context, and is thus less constrained. A neural net with the output classes $C_{NN} = \{silence, ivcorrect, ivincorrect, oov\}$ is used to determine

- whether a recognized word is overlapped with OOV speech or not
- whether a word is mis-recognized or not
- whether a word is OOV or not, given the word was mis-recognized
- what is the most probable class $c \in C_{NN}$ of the word

In [Kom09] we applied this approach to noisy telephone speech, reported improvements, and found it to generalize reasonably well across different data sets.

3   OOV Word Detection Based on a Hybrid Word/Sub-word Recognizer

However, our NN-based OOV detection approach does not retrieve a description of the underlying OOV, and, in cases where the recognized word boundaries do not match the reference, it cannot indicate accurately where, within a word, an OOV starts or ends. That is why we recently used a hybrid recognizer which consists of specific word models and a generic word model [Del95] for OOV word detection. The generic model is able to detect OOV words as sequences of sub-words. The search for the most likely word sequence can choose either an in-vocabulary word or the generic word as shown in

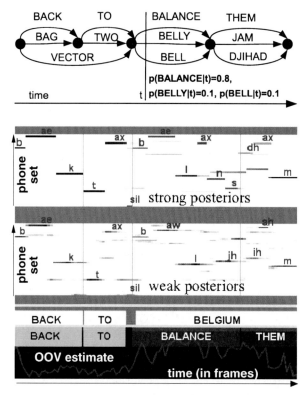

Fig. 1 NN-based system detecting the OOV word "BELGIUM": Processing of a word lattice (top) produced by the specific recognizer. The incongruence
between the specific and the generic phone posteriors is detected by the neural net
and identifies the corresponding words as OOVs (bottom).

figure 2. We compare the real output of an existing word-only recognizer and
the best possible output of a hybrid word/sub-word recognizer, respectively:

```
 reference: SORT OF A BLUEISH(OOV) MEDIUM
 word rec: SORT OF EVOLUTION        YOU
hybrid rec: SORT OF A bl.uw.ih.sh  MEDIUM
```

It can be seen, that the hybrid recognizer carries potential to simplify and
improve the detection and localization of OOV words over our NN-based
system. This is mainly due to the following reasons:

- The resulting word boundaries in OOV regions are more flexible, thus
  potentially more accurate. Context words are less often mis-recognized.
- The decision of the recognizer to prefer sub-word sequences over word
  sequences provides good evidence for an OOV word.

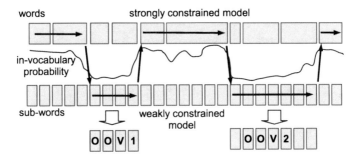

Fig. 2 OOV word detection using hybrid recognition: The best path (arrows) contains words and sub-word sequences, which can be regarded as potential OOVs. In addition, a frame-wise confidence measure is extracted from the combined word/sub-word lattice output of the recognizer shown as in-vocabulary probability.

- Often, two or three words in the word recognition are overlapped with a single OOV word. When using the hybrid recognition output, however, in many cases one sub-word sequence aligns to just one reference OOV word.

Using this setup, we have two possible choices for evaluation: Either we treat each sub-word sequence in the recognition output as potential OOV. This yields high precision, but many OOV words are missed. Alternatively, all words and sub-word sequences in the recognition output can be potential OOVs, which corresponds to the task performed previously using the neural-net based OOV detection system. In that case the recall in OOV detection improves, but the number of false alarms increases and the regions of OOV words tend to be less accurate. In case the detected OOV word was decoded as a sub-word sequence, we implicitly obtained a phonetic description of the OOV. Unlike before, we now just performed OOV detection using a hybrid confidence measure estimating the posterior probability for $C_{hybrid} = \{iv, oov\}$.

4   Fusion of Both Methods

We combined the scores of our both OOV detection methods by using linear logistic regression. 2.5 hours of Fisher data (telephone speech) were used for training and 7.5 hours for evaluation. The OOV rate was around 6.1%, and the neural-net based OOV detection system was trained using a disjunctive set of OOV words. All scores for the fusion were created initially on frame-level[1] and represented posterior probabilities:

$$\sum_{c \in C} p(c|frame) = 1, \ C \in \{C_{NN}, C_{hybrid}\} \tag{1}$$

---

[1] 10 ms length.

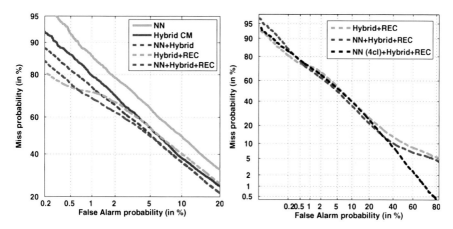

Fig. 3 Combined OOV word detection performance: Detection error trade-off across a wide range (right) and a range suitable for real application (left).

A hybrid confidence measure (Hybrid CM) estimating a probability of being in OOV was extracted from the lattice output of the hybrid recognizer and a binary score (REC) based on the recognition output of the hybrid recognizer (1 for frames covering sub-word sequences, 0 otherwise) were included in the fusion experiment. Our NN-based system estimated posterior probabilities of four classes using two neural nets using different type of context in the input [Kom09]. We converted the posterior probabilities into log-likelihood ratios[2] and averaged them over the word boundaries provided by the hybrid recognition output to obtain word-level scores.

Figure 4 shows OOV word detection performance of scores of both systems (bold lines) and their fusion (dashed lines). The left plot shows the zoomed view of the operational range reasonable for almost all tasks. The performance of all scores across a wide range is shown in the right plot. The best performance is achieved using different fusions for different ranges of false alarms (FA):

1. Up to 0.57% FA - hybrid system only
2. From 0.57% up to 20% FA - hybrid and NN system
3. From 20% FA - hybrid and NN(4cl) system

In the first range, we obtain a high precision in OOV detection. The best fusion intersects with the operation point determined by the binary score obtained from the word/sub-word recognition output. This is around 0.57% FA, where the fusion during the *second range* slowly starts to gain from the NN-based scores. Here, we retrieve already more OOV targets as opposed to the smaller amount of targets contained in the sub-word sequences in the word/sub-word recognition output of the hybrid system. *The third range*

---

[2] $LLR(p) = \ln \frac{p}{1-p}$.

benefits from using the scores of all four classes of the neural net. Some OOV words gets detected better by the NN-based system, but at the cost of retrieving many false alarms - far too many to be of practical use.

To conclude, the NN-based score improves the OOV detection performance across a wide range when fused with the hybrid CM. However, the better decision is to use the one-best binary score in the fusion, unless recall is more important than precision. In that case, the neural net is still able to retrieve some OOV words which otherwise would have been missed in the mid-range of the detection error trade-off curve.

5   Beyond OOV Detection

Upon detecting an unexpected event, the system should react. As a default strategy, even ignoring words detected as OOVs prevents mis-recognitions. However, unexpected events potentially carry a high amount of information - i.e. OOVs are most often content words. Thus, it is desirable to localize and analyze the event, which is a prerequisite for further processing stages to deal with the event in a more sophisticated way. The following actions could be taken upon detection of an OOV word:

- *Analysis:* obtain a phonetic description.
- *Recovery:* obtain the spelling and insert it into the recognizer output.
- *Judge importance:* some classes of OOVs might be particularly interesting, e.g. the class of OOVs that suddenly occur several times, such as the name of a formerly unknown politician in broadcast news.
- *Query-by-example:* find other examples of the same word.
- *Similarity scoring:* group re-occurring (or similar) unknown words.
- *Higher level description:* relate the new word to known words and to other detected OOVs.
- *Model update:* estimate a new word model and integrate it into the system.

The usefulness of particular OOV detections may vary from task to task. If it is just to detect mis-recognized words in the recognition output (due to the presence of an OOV), it is sufficient to find a single phone or frame in the word that has a low confidence score. However, if the task is to describe the OOV or to retrieve other examples of it, detecting a single phone of the OOV is not any helpful - we need to get the OOV region as exactly as possible. Therefore, we analyze detections by measuring recall of the OOV region and precision of the detected region (the sub-word sequence), as shown in figure 4.

5.1   Spelling Recovery of OOVs

Using grapheme-to-phoneme (G2P) conversion [Bis08], we retrieve the spelling of a word from the phonetic description. By substituting the sub-word sequences with the estimated spellings, we are able to correct a significant

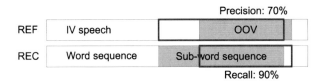

Fig. 4 The quality of a detected OOV word is determined by precision and recall.

portion of recognition errors due to OOVs [Kom10] and can also identify false alarms, in case the sub-word sequences convert back to known words. The retrieved spelling is a human readable representation of the OOV (e.g. EXTINCTION, PANDEMIC, GRAVITATIONAL) , which is interpretable within context also in case of slight errors (e.g. COURTICAL, EMBALLISHMENT).

5.2 Finding Re-occurring OOVs

Due to the higher level structure of audio/texts (into documents, broadcast shows, telephone calls), several OOVs do not only occur once, but repeat several times within different contexts. Those words often belong to topic-related vocabulary and are particularly important. Given one example of the word, we want to find other examples (query-by-example) and we want to cluster all detected OOVs to judge, whether some of them are re-occurring, and thus, important. For both tasks, we need a similarity measure of detected OOVs. The phonetic description of the detected OOVs, however, will not match precisely, as shown in this example detections for the OOV "IL-LUMINATION":

```
ax l uw m ax n ey sh en
   l ih m ax n ey sh en z
```

In [Han10], we described a similarity measure based on the alignment of recognized sub-word sequences. With the help of an alignment error model, which is able to deal with recognition errors and boundary mismatches (varying recall and precision of OOV region), we could retrieve roughly 60% of the re-occurring OOVs in telephone calls.

5.3 Relating OOVs to Other Words

Looking at examples of OOVs [Han10], we observe that unknown words most often are not entirely unknown. Except e.g. proper names in foreign languages, the majority of OOVs can be - morphologically or semantically - related to other known words or to other OOVs (derivational suffixes, semantic prefixes, compound words). Such a higher-level description of the

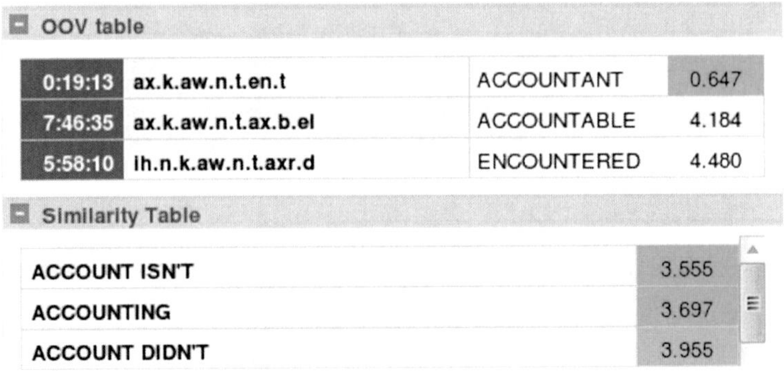

OOV table			
0:19:13	ax.k.aw.n.t.en.t	ACCOUNTANT	0.647
7:46:35	ax.k.aw.n.t.ax.b.el	ACCOUNTABLE	4.184
5:58:10	ih.n.k.aw.n.t.axr.d	ENCOUNTERED	4.480

Similarity Table	
ACCOUNT ISN'T	3.555
ACCOUNTING	3.697
ACCOUNT DIDN'T	3.955

Fig. 5 OOV demo on the selected OOV detection 'ax.k.aw.n.t.en.t': the top table shows time stamps where similar detections are found, and their recovered spelling, respectively. The output is ranked by a similarity score, with the selected detection ranking at top. The bottom table shows similar IV/OOV compounds.

unknown word can identify word families and identify the parts of the word, that are not modeled yet. We achieved this analysis by aligning a detected OOV to sequences of IVs and other detected OOVs. This is essentially a second stage of decoding, where we decode the detected sub-word sequences using a vocabulary consisting of all IV words and all other detected OOVs.

Figure 5 shows a screen shot of our OOV word detection and recovery demo available at http://www.lectures.cz/oov-fisher. It demonstrates the follow-up tasks such as spelling recovery, finding of similar OOV detections using similarity scoring and related compounds created out of known and unknown words.

6 Conclusions

In this work, we investigated into two approaches for OOV word detection. We compare both systems in a fusion experiment, and describe how to actually make use of the detected incongruence. We successfully implemented some out of the proposed follow-up actions (spelling recovery, similarity scoring and higher level description). Our approach relates parts which are well-known (sub-word units) to whole words which are not modelled yet (OOV words), which corresponds to the part-membership relationship postulated in the theoretical DIRAC framework.

Speech recognition is a sequential problem: prevention of damage in the context, and identification of the region affected by an unexpected event is particularly important to us. When aiming to go beyond OOV word detection, it became clear, that designing a system just primarily for detecting unexpected events might not be desirable. This became clear, especially when specific and generic systems were combined for the purpose of incongruence detection, but the localization was difficult and valuable information necessary for the follow-up process was lost. After extending our first approach by a hybrid recognition, we improve detection, and sustain higher accuracy in localization.

Another conclusion is, that a standard task definition for OOV word detection does not exist, and neither does it seem reasonable to define it. The usefulness of a particular OOV detection depends highly on the intended follow-up tasks, which again commends to first examine *how to react* on an unexpected event, in order to gain insights about how to improve its detection.

References

[Del95]   Deligne, et al.: Language Modeling by Variable Length Sequences: Theoretical Formulation and Evaluation of Multigrams. In: ICASSP, Detroit, MI, pp. 169–172 (1995)

[Jia05]   Jiang, H.: Confidence measures for speech recognition: A survey. Speech communication 45(4), 455–470 (2005)

[Bis08]   Bisani, M., Ney, H.: Joint-sequence models for grapheme-to-phoneme conversion. Speech Communication 50(5), 434–451 (2008)

[Kom09]   Kombrink, S., Burget, L., Matějka, P., Karafiát, M., Heřmansky, H.: Posterior-based Out-of-Vocabulary Word Detection in Telephone Speech. In: Proc. Interspeech 2009, Brighton, UK (2009)

[Han10]   Hannemann, M., Kombrink, S., Burget, L.: Similarity Scoring for Recognizing Repeated Out-of-Vocabulary Words. Submitted to Interspeech, Tokyo, JP (2010)

[Kom10]   Kombrink, S., Hannemann, M., Burget, L., Heřmansky, H.: Recovery of rare words in lecture speech. Accepted for Text, Speech and Dialogue (TSD), Brno, CZ (2010)

Incongruence Detection in Audio-Visual Processing

Michal Havlena, Jan Heller, Hendrik Kayser, Jörg-Hendrik Bach,
Jörn Anemüller, and Tomáš Pajdla

Abstract. The recently introduced theory of incongruence allows for detection of unexpected events in observations via disagreement of classifiers on specific and general levels of a classifier hierarchy which encodes the understanding a machine currently has of the world. We present an application of this theory, a hierarchy of classifiers describing an audio-visual speaker detector, and show successful incongruence detection on sequences acquired by a static as well as by a moving AWEAR 2.0 device using the presented classifier hierarchy.

1 Theory of Incongruence

The recently introduced theory of incongruence [6, 9] allows for detection of unexpected events in observations via disagreement of classifiers on specific and general levels of a classifier hierarchy which encodes the understanding a machine currently has of the world. According to [6, 9], there are two possibilities of how an incongruence can appear. In a class-membership hierarchy, incongruence appears when a general classifier accepts but all the specific, i.e. child, classifiers reject. This means that possibly a novel sub-class of objects has been observed. On the other hand, in a part-whole hierarchy, incongruence appears when all general, i.e. parent, classifiers accept but a specific classifier does not. This is often caused by the fact that the hierarchy is incomplete, some of the classifiers on the general level of the hierarchy may be missing.

Michal Havlena · Jan Heller · Tomáš Pajdla
Center for Machine Perception, Department of Cybernetics, FEE, CTU in Prague,
Technická 2, 166 27 Prague 6, Czech Republic
e-mail: {havlem1,hellej1,pajdla}@cmp.felk.cvut.cz

Hendrik Kayser · Jörg-Hendrik Bach · Jörn Anemüller
Medizinische Physik, Fakultät V, Carl von Ossietzky - Universität Oldenburg,
D-26111 Oldenburg, Germany
e-mail: {hendrik.kayser,j.bach}@uni-oldenburg.de,
    joern.anemueller@uni-oldenburg.de

D. Weinshall, J. Anemüller, and L. van Gool (Eds.): DIRAC, SCI 384, pp. 67–75.

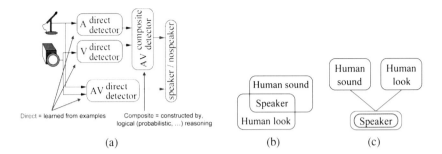

(a)   (b)   (c)

Fig. 1 (a) "Speaker" event can be recognized in two ways, either by a holistic (direct) classifier, which is trained directly from complete audio-visual data, or by a composite classifier, which evaluates the conjunction of direct classifiers for "Human sound" and "Human look" events. (b) "Speaker" is given by the intersection of "Human sound" and "Human look". (c) "Speaker" corresponds to the infimum in the Boolean POSET.

Our audio-visual speaker detector is an example of a part-whole hierarchy. Figure 1 shows "Speaker" event that is recognized in two ways, either by the direct classifier, which is trained directly from complete audio-visual data, or by the composite classifier that evaluates the conjunction of direct classifiers for "Human sound" and "Human look" events. "Speaker" is given by the intersection of sets representing "Human sound" and "Human look" which corresponds to the infimum in the Boolean POSET. In the language of [6, 9], the composite classifier corresponds to the general level, i.e. to $Q^g_{speaker}$, while the direct classifier corresponds to the specific level, i.e. to $Q_{speaker}$.

The direct audio classifier, see Figure 2(a), detects human sound, e.g. speech, and returns a boolean decision on "Human sound" event. The direct visual classifier, see Figure 2(b), detects human body shape in an image and returns a boolean decision on "Human look" event. The direct audio-visual classifier, see Figure 3(a), detects the presence of a speaker and returns a boolean decision on "Speaker" event. The composite audio-visual classifier, see Figure 3(b), constructed as the conjunction of the direct audio and visual classifiers also detects the presence of a speaker and returns a boolean decision on "Speaker" event. Opposed to the direct audio-visual classifier, its decisions are constructed from the decisions of the separate classifiers using logical AND. After presenting a scene with a silent person and speaking loudspeaker, the composite audio-visual classifier accepts but the direct audio-visual classifier does not. That creates a disagreement, incongruence, between classifiers. Table 1 interprets the results of "Speaker" event detection.

According to the theory, collected incongruent observations can be used to refine the definition of a speaker used by the machine, i.e. to correct the understanding of the world, by adding a classifier to the general level of the hierarchy. The composite audio-visual classifier of the refined "Speaker" event detector will not accept scenes with silent people and speaking loudspeakers because they will be rejected by the added classifier. As the direct audio-visual classifier also rejects such scenes, no

Table 1 Interpretation of agreement/disagreement for "Speaker" event example.

	Composite Q_a^g (general level)	Direct Q_a (specific level)	Possible reason
1	$0 \simeq Q_a^g \simeq Q_a \simeq 0$ reject	reject	empty silent scene
2	$1 \simeq Q_a^g \gg Q_a \simeq 0$ accept	reject	silent person & speaking loudspeaker
3	$0 \simeq Q_a^g \ll Q_a \simeq 1$ reject	accept	inconsistent POSET, wrong model
4	$1 \simeq Q_a^g \simeq Q_a \simeq 1$ accept	accept	speaking person

incongruence will be rendered. The detailed description of the refinement of the definition of a speaker, which is beyond the scope of this paper, can be found in [5] or in an accompanied paper.

2 Audio-Visual Speaker Detector

Next, we describe the classifiers used in deeper detail.

2.1 Direct Audio Detector

The direct audio classifier for sound source localization is based on the generalized cross correlation (GCC) function [4] of two audio input signals. A sound source is localized by estimating the azimuthal angle of the direction of arrival (DOA) relative to the sensor plane defined by two fronto-parallel microphones, see Figure 2(a). The approach used here enables the simultaneous localization of more than one sound source in each time-frame.

The GCC is an extension of the cross power spectral density function, which is given by the Fourier transform of the cross correlation. Given two signals $x_1(n)$ and $x_2(n)$, it is defined as:

$$G(n) = \frac{1}{2\pi} \int_{-\infty}^{\infty} H_1(\omega) H_2^*(\omega) \cdot X_1(\omega) X_2^*(\omega) e^{j\omega n} \, d\omega, \tag{1}$$

where $X_1(\omega)$ and $X_2(\omega)$ are the Fourier transforms of the respective signals and the term $H_1(\omega) H_2^*(\omega)$ denotes a general frequency weighting. We use PHAse Transform (PHAT) weighting [4], which normalizes the amplitudes of the input signals to unity in each frequency band:

$$G_{PHAT}(n) = \frac{1}{2\pi} \int_{-\infty}^{\infty} \frac{X_1(\omega) X_2^*(\omega)}{|X_1(\omega) X_2^*(\omega)|} e^{j\omega n} \, d\omega, \tag{2}$$

such that only the phase difference between the input signals is preserved.

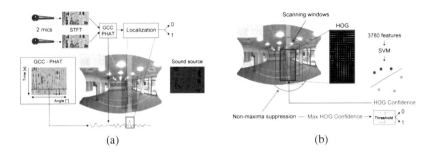

(a)   (b)

Fig. 2 (a) The direct audio detector employing GCC-PHAT. A sound source is localized by estimating the azimuthal angle of the direction of arrival (DOA) relative to the sensor plane. (b) The direct visual detector based on work [2].

From the GCC data, activity of a sound source is detected for 61 different DOA angles. The corresponding time delays between the two microphones cover the field of view homogeneously with a resolution of 0.333 ms. The mapping from the time delay τ to the angle of incidence θ is non-linear:

$$\theta = \arcsin\left(\tau \cdot \frac{c}{d}\right), \tag{3}$$

where c denotes the speed of sound and d the distance between the sensors. This results in a non-homogenous angular resolution in the DOA-angle space, with higher resolution near the center and lower resolution towards the edges of the field of view. If at least in one of these 61 directions a sound source is detected, the direct audio classifier accepts.

2.2 Direct Visual Detector

The state-of-the-art paradigm of visual human detection [2] is to scan an image using a sliding detection window technique to detect the presence or absence of a human-like shape. By adopting the assumption that the ground plane is parallel to the image, which is true for our static camera scenario, we can restrict the detection window shapes to those that correspond to reasonably tall (150–190 cm) and wide (50–80 cm) pedestrians standing on the ground plane. Visual features can be computed in each such window as described below.

For the direct visual classifier, we use INRIA OLT detector toolkit [1], based on the histograms of oriented gradients (HOG) algorithm presented by Dalal and Triggs [2]. It uses a dense grid superimposed over a detection window to produce a 3,780 dimensional feature vector. A detection window of 64×128 pixels is divided into cells of 8×8 pixels, each group of 2×2 cells is then integrated into overlapping blocks. Each cell consists of 9-bin HOG that are concatenated for each block and normalized using L_2 norm. Each detection window consists of 7×15 blocks resulting in 3,780 features. As the cylindrical projection used to represent image data in our experiments is locally similar to the perspective projection, we used a detector trained on perspective images.

An image is scanned by the detection window at several scales and positions in a grid pattern. Detection scores (confidences) higher than 0, i.e. tentative human detections, are further processed using non-maxima suppression with robust mean shift mode detection, see Figure 2(b). If the detection score of the best detection is higher than 0.1, the direct visual classifier accepts.

2.3   Direct Audio-Visual Detector

For the direct audio-visual classifier, we use the concept of angular (azimuthal) bins that allows for handling multiple pedestrians and/or sound sources at the same time. $180°$ field of view is divided into twenty bins, each $9°$ wide, with the classification performed per bin. If at least one of the bins is classified as positive, the direct audio-visual classifier accepts. The detailed procedure of the classification of a single bin follows.

First, a 2D feature vector is constructed from audio and visual features as the highest GCC-PHAT value and the highest pedestrian detection score belonging to the bin. The pedestrian detection score is maximized both in the x and y coordinates of the window center where x has to lie in the bin and y goes through the whole height of the image as the bins are azimuthal, i.e. not bounded in the vertical direction. Then, the feature vector is classified by SVM with the RBF kernel [7]. Any non-negative SVM score yields a positive classification.

The SVM classifier was trained using four sequences (2,541 frames in total) of people speaking while walking along a straight line. Every frame of the training sequences contains one manually labeled positive bin and the 2D feature vectors corresponding to these bins were taken as positive examples. The two bins adjacent to the labeled ones were excluded from training and the feature vectors corresponding to the rest of the bins were taken as negative examples. As examples of people not speaking and of speech without people, which were not in the training data, were also needed, we created negative examples for these cases by combining parts of the positive and negative feature vectors, yielding 109,023 negative examples in total. The training data together with the trained decision curve can be seen in Figure 3(a).

2.4   Composite Audio-Visual Detector

The composite audio-visual classifier was constructed to explain audio-visual events by a combination of simpler audio and visual events. At the moment, the classifiers are combined using a two-state logic, which is too simplistic to cope with the full complexity of real situations but it provides all basic elements of reasoning that later could be modeled probabilistically. Opposed to the direct audio-visual classifier, which separates relevant events from the irrelevant ones, the composite audio-visual classifier represents the understanding a machine currently has of the world.

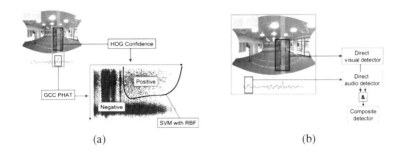

(a)   (b)

Fig. 3 (a) The direct audio-visual detector constructed from audio and visual features. Training data and the trained decision curve for SVM with RBF of the direct audio-visual classifier: GCC-PHAT values (*x*-axis) and pedestrian detection scores (*y*-axis) for different positive (*red circles*) and negative (*blue crosses*) manually labeled examples. (b) The composite audio-visual detector constructed as the conjunction of the decisions of the direct audio and visual detectors.

The composite classifier is not meant to be perfect. It is constructed as the conjunction of the decisions of the direct audio and visual classifiers, each evaluated over the whole image. Hence, it does not capture spatial co-location of the audio-visual events. It detects events concurrent in time but not necessarily co-located in the field of view, Figure 3(b). It will fire on sound coming from a loudspeaker on the left and a silent person standing on the right when that happens at the same moment. This insufficiency is intentional. It models current understanding of the world, which is due to its complexity always only partial, and leads to detection of incongruence w.r.t. the direct audio-visual classifier when human sound and look come from dislocated places.

3   Experimental Results

The direct and composite audio-visual classifiers were used to process several sequences acquired by the AWEAR 2.0 device [3], comprised of 3 computers, 2 cameras with fish-eye lenses heading forwards, and a Firewire audio capturing device with 2 forward and 2 backward-heading microphones. The device can be placed on a table or worn as a backpack thanks to 4 lead-gel batteries which ensure the autonomy of about 3 hours. Audio and visual data are synchronized using a hardware trigger signal.

3.1   Static Device

The 30 s long sequence SPEAKER&LOUDSPEAKER shot at 14 fps acquired by the static device contains a person speaking while walking along a roughly straight line. After a while the person stops talking and the loudspeaker sounds up, which renders an incongruent observation and should be detected.

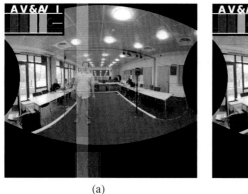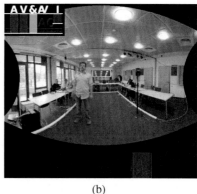

(a)   (b)

Fig. 4 Two example frames from SPEAKER&LOUDSPEAKER sequence showing (a) a congruent and (b) an incongruent situation when the speaker and the loudspeaker sound respectively. Decisions from different direct classifiers are drawn into the frames: direct audio classifier (*magenta*), direct visual classifier (*blue*), direct audio-visual classifier (*green*). The bars in the top-left corner show also the composite audio-visual classifier decisions (*cyan*) and incongruence/wrong model (*red/yellow*).

As the video data come from the left camera of a stereo rig with a 45 cm wide baseline, there is a discrepancy between the camera position and the apparent position of a virtual listener, located at the center of the acquisition platform, to which the GCC-PHAT is computed. To compensate for this, the distance to the sound source and the distance between the virtual listener and the camera need to be known. The listener–camera distance is 22.5 cm and can be measured from the known setup of the rig. The distance to the sound source is assumed to be 1.5 m from the camera. The corrected angle can be then trivially computed from the camera–listener–sound source triangle using a line–circle intersection. Due to this angle correction, the accuracy of incongruence detection is lower for speakers much further away than the assumed 1.5 m. Two example frames from the sequence can be found in Figure 4 together with the decisions from different classifiers.

3.2 Moving Observer

The script of the 73 s long 14 fps sequence MOVINGOBSERVER is similar to the one of SPEAKER&LOUDSPEAKER sequence, the difference being the fact that the AWEAR 2.0 device was moving and slightly rocking when acquiring it. As the microphones are rigidly connected with the cameras, the relative audio-visual configuration remains the same and movement does not cause any problems to the combination of the classifiers. On the other hand, the direct visual classifier is able to detect upright pedestrians only, therefore we perform pedestrian detection in frames stabilized w.r.t. the ground-plane [8] and transfer the results back into the original frames yielding non-rectangular pedestrian detections, see Figure 5.

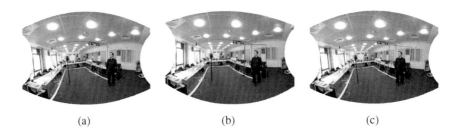

(a)   (b)   (c)

Fig. 5 Pedestrian detection in frames stabilized w.r.t. the ground-plane transferred back to the original frame. (a) Non-stabilized frame. (b) Stabilized frame with pedestrian detection. (c) Non-stabilized frame with the transferred pedestrian detection. Notice the non-rectangular shape of the detection window after the transfer.

As the x coordinate of the mouth can be different than the x coordinate of the center of the detected window for non-upright pedestrians, we use the position of the center of the upper third of the window (i.e. the estimated mouth position) in the direct audio-visual classifier instead.

4   Conclusions

We presented an application of the theory of incongruence, a hierarchy of classifiers describing an audio-visual speaker detector, and showed successful incongruence detection on sequences acquired by the AWEAR 2.0 device. Collected incongruent observations can be used to refine the definition of a speaker used by the machine, i.e. to correct the understanding of the world, by adding a classifier to the general level of the hierarchy, see [5] or an accompanied paper for details.

Acknowledgements. This work was supported by the EC project FP6-IST-027787 DIRAC and by Czech Government under the research program MSM6840770038. Any opinions expressed in this paper do not necessarily reflect the views of the European Community. The Community is not liable for any use that may be made of the information contained herein.

References

1. Dalal, N.: INRIA Object Detection and Localization Toolkit (2008), Software,
   http://pascal.inrialpes.fr/soft/olt
2. Dalal, N., Triggs, B.: Histograms of oriented gradients for human detection. In: CVPR 2005, vol. 2, pp. 886–893 (2005)
3. Havlena, M., Ess, A., Moreau, W., Torii, A., Jančošek, M., Pajdla, T., Van Gool, L.: AWEAR 2.0 system: Omni-directional audio-visual data acquisition and processing. In: EGOVIS 2009: First Workshop on Egocentric Vision, pp. 49–56 (2009)
4. Knapp, C., Carter, G.: The generalized correlation method for estimation of time delay. IEEE Transactions on Acoustics, Speech and Signal Processing 24(4), 320–327 (1976)

5. Pajdla, T., Havlena, M., Heller, J., Kayser, H., Bach, J.H., Anemüller, J.: Incongruence detection for detecting, removing, and repairing incorrect functionality in low-level processing. Research Report CTU–CMP–2009–19, Center for Machine Perception, K13133 FEE Czech Technical University (2009)
6. Pavel, M., Jimison, H., Weinshall, D., Zweig, A., Ohl, F., Hermansky, H.: Detection and identification of rare incongruent events in cognitive and engineering systems. Dirac white paper, OHSU (2008)
7. Schölkopf, B., Smola, A.: Learning with Kernels. The MIT Press, MA (2002)
8. Torii, A., Havlena, M., Pajdla, T.: Omnidirectional image stabilization by computing camera trajectory. In: Wada, T., Huang, F., Lin, S. (eds.) PSIVT 2009. LNCS, vol. 5414, pp. 71–82. Springer, Heidelberg (2009)
9. Weinshall, D., et al.: Beyond novelty detection: Incongruent events, when general and specific classifiers disagree. In: NIPS 2008, pp. 1745–1752 (2008)

Catalog of Basic Scenes for Rare/Incongruent Event Detection

Danilo Hollosi, Stefan Wabnik, Stephan Gerlach, and Steffen Kortlang

Abstract. A catalog of basic audio-visual recordings containing rare and incongruent events for security and in-home-care scenarios for European research project *Detection and Identification of Rare Audio-visual Cues* is presented in this paper. The purpose of this catalog is to provide a basic and atomistic testbed to the scientific community in order to validate methods for rare and incongruent event detection. The recording equipment and setup is defined to minimize the influence of error that might affect the overall quality of the recordings in a negative way. Additional metadata, such as a defined format for scene descriptions, comments, labels and physical parameters of the recording setup is presented as a basis for evaluation of the utilized multimodal detectors, classifiers and combined methods for rare and incongruent event detection. The recordings presented in this work are available online on the DIRAC preoject website [1].

1 Introduction

The performance of state-of-the-art event detection schemes usually depend on a huge amount of training data to be able to identify an event correctly. The less data is available, the more problematic the abstraction of the event to a model becomes. This is especially true for rare event detection, where events have a very low a-priori probability. In the same context, this is also true for incongruent events. They are defined as rare events which conflict with commonly accepted world models. Consequently, novel methods need to be developed to detect such events and to compensate for the lack of suitable training data.

Danilo Hollosi · Stefan Wabnik
Fraunhofer Institute for Digital Media Technology, Project Group Hearing Speech and Audio Technology, D-26129 Oldenburg, Germany
e-mail: Danilo.Hollosi@idmt.fraunhofer.de

Stephan Gerlach · Steffen Kortlang
Carl von Ossietzky University of Oldenburg,
Institute of Physics, D-26111 Oldenburg, Germany

D. Weinshall, J. Anemüller, and L. van Gool (Eds.): DIRAC, SCI 384, pp. 77–84.
springerlink.com

Possible solutions for this problem are currently investigated in the European research project *Detection and Identification Of Rare Audio-visual Cues* (DIRAC). The main idea of the DIRAC approach for event detection is - in contrast to existing holistic models - to make use of the discrepancy between more general and more specific information available about reality [2]. This approach heavily reduces the amount of training data necessary to model rare and incongruent events and can be generalized in such a way that information from different modalities become applicable as well.

Two application domains are defined for the DIRAC project, namely the security market with its high demand for automated and intelligent surveillance systems [3], and the in-home care market with its need for monitoring elderly people at home [4]. It was concluded that both domains would benefit considerably from the technology developed in the DIRAC project. In both domains there is a need for 24/7, unobtrusive, autonomous, and therefore intelligent, monitoring systems to assist human observers. Spotting and properly responding to unforeseen situations and events is one of the crucial aspects of monitoring systems in both application domains.

For both of them, scenarios have been developed and example situations have been recorded to show the potential of the DIRAC theoretical framework, while attempting to address realistic and interesting situations that can not be handled properly by existing technology. The methods show promising results on first recordings of near-real-life situations, but additional data is necessary to provide a strategy for testing and validation of those methods within the different scenarios of an application domain.

This paper presents a catalog of recordings of very basic, atomistic scenes containing incongruent events that are suitable for testing and validation of the DIRAC methods. Very basic in this context means for example only one person, only one action/incongruency, indoors, no cast shadows, enough light and defined sound sources. To support the idea of a controlled environment, the scenes and the necessary restrictions are defined in detail. Thus, it will be easier to compose more complex scenarios from atomistic scenes without having too much uncertainty about what can be processed with the underlying, utilized detectors and classifiers forming the general and specific models within a DIRAC method.

This work is organized as follows. First of all, the recording equipment and set is defined in detail in order to avoid errors and artifacts which would reasonably lead to unwanted results in individual classifiers and DIRAC methods. This includes a description of the AWEAR-II recording system in section 2.1, information on the Communication Acoustic Simulator (CAS) recording room in section 2.2 and the recording setup itself in section 2.3. Afterwards, potential sources of errors are investigated and used to define characteristics of suitable audio-visual recordings in section 2.4. To the heart of this work, the catalog of basic audio-visual scenes is presented in section 3, followed by information on the format of the audio-visual recordings, their labels, scene descriptions and additional metadata in section 3.1 and 3.2.

2 Methods

2.1 AWEAR-II Recording System

The AWEAR-II recording platform is a portable audio-visual recording platform that was developed in the DIRAC project. It consists of three mini-pcs (*Siemens D2703-S mini ITX boards*), two cameras (*AVT Stingray*), four microphones (*Tbone EM700 stereo mic set*), an audio capturing interface including a triggering device for audio-visual synchronization (*FOCUSRITE Saffire PRO 10*), a battery pack (*Camden 12V gel*) and a power distribution box. The hardware is mounted on a wearable backpack frame and allows human-centered recordings in both indoor and outdoor environments [5]. A picture of the AWEAR-II recoding system can be found in Fig. 1.

The AWEAR-II system can be controlled using a graphical user interface (GUI) application running on a netbook which in turn communicates with the recording platform via a wireless network connection. The GUI is used as a remote control for the preparation of recording sessions and capturing, for hardware adjustments and controlling. The recording data is stored on 2,5" removable hard disks connected to the AWEAR-II. After a recording session, the data is copied from the disks to a dedicated PC for further processing. For this purpose, format conventions have been defined will be described in section 3.1 and section 3.2.

2.2 Communication and Acoustics Simulator - CAS

The Communication and Acoustics Simulator (CAS) at the *House of Hearing* in Oldenburg is a special room with variable acoustics that uses sophisticated techniques consisting of countless microphones, loudspeakers as well as large-scale electronics to create almost any acoustic room condition desired [6]. The CAS is usually used to run subjective tests on the latest hearing aids under various acoustic conditions or to test mobile phones for speech recognition and intelligibility in realistic environments, but is also interesting for our work since environment dependent parameters can be manually controlled.

For part of the recordings, the variable acoustics of the CAS was set up to the *living room* scenario with a reverberation time $T60$ of around 0.453 seconds. A floor-plan of the CAS can be found in Fig. 2, whereas additional information about the camera location and walking paths can be found there as well.

2.3 Recording Setup

In this section, the scenes and the necessary restrictions are defined in more detail. The actors use a defined path to enter the scenery, to go trough the scenery and to leave it. Therefore, way points and walking paths have been defined and marked on the floor of the CAS in order to ensure repeatability of the experiments and and recordings. The AWEAR-II was placed orthogonal to the back wall in a distance of

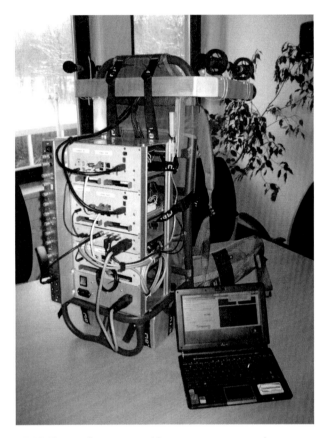

Fig. 1 The AWEAR-II recording system with nettop remote control

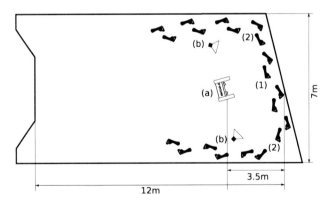

Fig. 2 Floor plan of the CAS and the proposed recording setup. *(a)* denotes the location of the AWEAR-II recording system, *(b)* is the location of the additional light sources, *(1)* is defined as the center of the cameras field of view at a distance of approximately two meters and *(2)* are additional markers for the actors walking paths.

three and a half meters, whereas the mentioned way points and walking paths are in parallel to the wall in the back with a distance of approximately two meters from the camera center. This distance was found to be suitable enough to capture a wide viewing angle. All the parameters have been summarized and illustrated in the CAS floor plan as it can be found in Fig. 2.

All the recordings were made during daytime. However, to be independent from weather situations, i.e. constant illumination, two additional light sources were added to the scene. The consist of 100W spots attached to a stand and a dimmer to control the amount of light introduced to the scene. Diffuse light situations are generated by using semi-transparent light shields which are attached to the spots.

2.4 Potential Sources of Errors

Providing high quality audio-visual recordings for testing and evaluation is a non-trivial task. There is more to it than to define an audio and video format and physical parameters, such as the video frame rate and the audio sampling rate for the recording setup only. The recording environment needs to be defined as well in order to provide the best possible audio-visual quality. Therefore, it is crucial to identify potential problems and artifacts beforehand in order to completely remove or at least minimize their influence on the quality.

The main sources of errors and artifacts in general can be found in the recording environment and in a wrong calibration of the equipment. In particular, problems with improper illumination and cast shadows are unnecessarily common as well as problems with foreground-background separation as a consequence of improper dressing of the actors. The consequences are misclassification and confusion when running video based detectors and classifiers. Blurring as a result of defocussing, wrong gamma settings, shutter and white imbalance are problems in this context as well.

For audio recordings, the presence of unwanted sound sources, reverberations, humming, noise introduction and too low recording levels (low SNR) are serious problem since they heavily influences the performance of audio based detectors and the overall quality of modality-fused detectors. Furthermore, improper synchronization between the audio and video data can be seen as a source of errors too.

2.5 Evaluation Criteria for Audio-Visual Recordings

Motivated from the information on potential errors and artifacts as described in the previous section, the following requirements and criteria to create basic and atomistic audio-visual scenes containing incongruent events have been developed. A *suitable* audio-visual sequence is defined as a sequence that allows error free processing of the DIRAC methods. Therefore, special attention needs to be drawn on sufficient illumination of the scene to minimize the influence of CCD camera chip noise and motion blurring artifacts, especially when working with low video frame rates and fast moving objects. Cast shadows should be avoided to reduce false results/alarms from the video detectors. Furthermore, a high contrast between foreground objects and a steady, homogeneous background is desirable in order to avoid misclassification and confusion within the automated processing stages. For the audio part, a suitable recording level should be selected such that the SNR stays reasonably high. Despite that the audio based detectors used in the DIRAC project have been shown to be robust against noise, the presence of uncontrollable noise sources should be minimized.

3 Basic Scenes for Incongruency Detection

Within the DIRAC project, keywords have been defined to describe the audio-visual recordings in the DIRAC in-home-care and security surveillance scenarios. This was done for two reasons: First, to allow a search for audio-visual scenes within a database based on the keywords and second, to form complexity scalable scenarios by combining keywords. At the moment, 6 keyword groups exist. They are Movements, Interactions, Audio and their corresponding incongruent versions. The content of the keyword groups can be seen in Table 1.

Table 1 Keyword groups, its members and incongruencies available in the catalog

Group	Members	Incongruency
Movements	standing	limping
	sitting	stumbling
	running	falling
	walking	backwards
	lying	sidewards
	hesitating	fleeing
Interactions	one to N persons	fighting
	person interact	
	dialog	
Audio	speech	monolog
	noise	shouting
		out-of-vocabulary

In total, 95 audio-visual scenes have been recorded, containing samples to test individual detectors and classifiers for moving objects, visual foreground and background detection, person detection and localization, voice activity detection, speech recognition and the detection of acoustic events. Furthermore, basic audio-visual scenes have been recorded which cover incongruencies and rare events such as the ones given with the keywords. Thus, a verification and evaluation of individual detectors and classifiers is possible as well as verification and evaluation of the more complex DIRAC methods. Of course, the recording are not only limited to the DIRAC methods, but can be used with any other fused modality approach.

3.1 Format of the Audio-Visual Recordings

In order to provide the best possible quality to the research community, all the recordings are stored in an uncompressed data format. In particular, the Portable Network Graphics (PNG) format is used camera channel-wise on a frame by frame basis with a 1920x1080 resolution. For the four microfon channels, all recordings are stored as wav-files with 48kHz sampling rate and a sample resolution of 32 bit float.

3.2 Metadata and Scence Descriptions

Each audio-visual recording contains a label file which includes information about the name of the recording and the scene, the date, the location, the used device, frame rate, a placeholder for comments, as well as a detailed description of the scene. All the labels have been generated manually, either by hand or by using custom made semi-supervised tools. Optionally, the audio-visual scene is rendered as a preview video using one or both camera signals and the front stereo microfon set. For this purpose, any video codec can be used since this is done only for preview purposes. In combination with the additional metadata and scene descriptions given in Fig. 4, selection of suitable audio-visual recordings is facilitated. The Metadata has been stored together with the audio-visual recordings on the DIRAC project page www.diracproject.org [1].

```
# Recording: AggressionRecordings
# Scene : AggressionScenes
# Date : 20091216
# Location : house front next to HdH (FRA Oldenburg)
# Equipment: AWEAR-II (fixed)
# Framerate: 12 fps
# Comments : shades visible

# start time (in sec) | end time (in sec) | key words | short description;
0000 | 0001 | pers. 2, walking | Two persons walk towards each other;
0002 | 0003 | pers. 2, walking, speech | Defensive person begins conversation;
0003 | 0005 | pers. 2, interact, shouting | Aggressor starts fighting;
0005 | 0010 | pers. 2, falling, fleeing | Person falls, aggressor flees;
0010 | 0017 | pers. 1, limping | Person limps away;
. . .
```

Fig. 3 Example of a label file to provide additional and contextual information about the audio-visual recording

4 Conclusion

A catalog of basic and atomistic audio-visual recordings for rare and incongruent event detection was presented in this paper. While adressing the work within the ongoing DIRAC project at first, the applicability of the catalog is not limited to the methods developed there. A careful definition and analysis of the recording environment, the recording equipment and setup is seen to be crucial for two reasons. First, to provide the best audio-visual quality of the recordings achievable to ensure that the performance of utilized detection schemes and classifiers do not degrade with quality of the test data. Second, to focus on the validation and evaluation of novel combined detection schemes and modality-fused event detection methods such as the ones proposed in DIRAC instead of the underlying algorithms for modeling only single information instances. The catalog will be used to validate the methods for rare and incongruent event detection developed within the DIRAC project and will probably be extended in the future based on the needs of the project partners. Both the recordings and the results will be published on the project website [1].

Acknowlegdement. The authors would like to thank *Haus des Hörens* for providing access to the CAS, technical support and equipment. This work was supported by the European Commission under the integrated project DIRAC (Detection and Identification of Rare Audio-visual Cues, IST-027787).

References

1. IST-027787 project website: Detection and Identification of Rare Audio-visual Cues - DIRAC, http://www.diracproject.org/
2. Weinshall, D., Hermansky, H., Zweig, A., Luo, J., Jimison, H., Ohl, F., Pavel, M.: Beyond Novelty Detection: Incongruent Events, when General and Specific Classifiers Disagree. In: Advances in Neural Information Processing Systems (NIPS), Vancouver (December 2008)
3. van Hengel, P.W.J., Andringa, T.C.: Verbal aggression detection in complex social environments. In: Proceedings of AVSS 2007 (2007)
4. van Hengel, P.W.J., Anemüller, J.: Audio Event Detection for In-Home-Care. In: NAG/DAGA International Conference on Acoustics, Rotterdam, March 23-26 (2009)
5. Havlena, M., Ess, A., Moreau, W., Torii, A., Janoek, M., Pajdla, T., Van Gool, L.: AWEAR 2.0 system: Omni-directional audio-visual data acquisition and processing. In: EGOVIS 2009: First Workshop on Egocentric Vision, pp. 49–56 (2009)
6. Behrens, T.: Der 'Kommunikationsakustik-Simulator' im Oldenburger Haus des Hörens. 31. Deutsche Jahrestagung für Akustik: Fortschritte der Akustik DAGA 2005, München, DEGA e.V., vol. (1), pp. 443–445 (2005)

Part III
Alternative Frameworks to Detect Meaningful Novel Events

Trajectory-Based Abnormality Categorization for Learning Route Patterns in Surveillance

Pau Baiget, Carles Fernández, Xavier Roca, and Jordi Gonzàlez

Abstract. The recognition of abnormal behaviors in video sequences has raised as a hot topic in video understanding research. Particularly, an important challenge resides on automatically detecting abnormality. However, there is no convention about the types of anomalies that training data should derive. In surveillance, these are typically detected when new observations differ substantially from observed, previously learned behavior models, which represent normality. This paper focuses on properly defining anomalies within trajectory analysis: we propose a hierarchical representation conformed by Soft, Intermediate, and Hard Anomaly, which are identified from the extent and nature of deviation from learned models. Towards this end, a novel Gaussian Mixture Model representation of learned route patterns creates a probabilistic map of the image plane, which is applied to detect and classify anomalies in real-time. Our method overcomes limitations of similar existing approaches, and performs correctly even when the tracking is affected by different sources of noise. The reliability of our approach is demonstrated experimentally.

1 Introduction

Recognizing abnormal behaviors is a main concern for research on video understanding [7]. The challenge exists not by the difficulty of implementing anomaly detectors, but because it is unclear how to generally define *anomaly*. On the one hand, anomalies in video surveillance are usually related to suspicious or dangerous behaviors, i.e., those for which an alarm should be fired. Unfortunately, such a vague concept is difficult to learn automatically without prior explicit formal models. Existing top-down approaches take advantage of this prior models to identify suspicious behavior [3] or to generate conceptual descriptions of detected behavior

Pau Baiget · Carles Fernández · Xavier Roca · Jordi Gonzàlez
Dept. Ciències de la Computació & Computer Vision Center
Edifici O, Campus Universitat Autnoma de Barcelona, 08193 Bellaterra, Barcelona, Spain
e-mail: {pbaiget,perno,xavir,poal}@cvc.uab.cat

D. Weinshall, J. Anemüller, and L. van Gool (Eds.): DIRAC, SCI 384, pp. 87–95.
springerlink.com

patterns [5]. However, these approaches are scenario oriented and are designed to recognize or describe a reduced set of specific behaviors.

On the other hand, from a statistical perspective, recent works in anomaly detection assume that normal behavior occurs more frequently than anomalous one, and define anomalies as deviations from what is considered to be normal [1, 8, 10, 11]. In surveillance, a standard learning procedure consists of extracting observations by motion tracking over continuous recordings to build scenario models that determine the normality or abnormality of new observations.

We define anomalies considering the types of deviations from a learned model. We distinguish among three semantic types, namely *Soft* (*SA*), *Intermediate* (*IA*), and *Hard* (*HA*) anomalies. In essence, a *SA* represents a slight deviation from the parameters of a typical pattern, e.g. a person running, stopping, walking backwards for a while, etc. Secondly, a *HA* occurs when an observed event occurs completely outside of the model parameters, e.g. a person appearing from an unnoticed entrance. These two types are currently recognized by different approaches, see Table 1, but there exists an important gap between them. The *IA* represents observations that deviate from learned patterns but still fit into the model, e.g., starting a typical path and changing to a different one.

To recognize these anomaly types, we present a novel unsupervised learning technique that generates a scenario model in terms of paths among observed entry and exit areas. We use Gaussian Mixture Models (GMM) to describe transition probabilities between pairs of entry and exit points. GMM model n-dimensional datasets, where each dimension is conditionally independent from the others. Thus, although our method currently uses only trajectory positions to fit GMM into the training set, it can be easily extended to features like bounding boxes, orientations, speed, etc. Additionally, our methodology identifies online previously introduced anomalies in new observed trajectories.

Recent surveillance techniques to learn motion patterns have been applied to anomaly detection, behavior description, semantic scene labeling, or tracking enhancement. Johnson and Hogg [9] and Fernyhough et al [6] use vector quantization to learn routes and semantic regions. The methods depends on target size. More recently, Makris et al [10] proposed a method to label semantic zones –entry, exit, stop area– and model typical routes. However, temporal consistency is not maintained, and speed variation anomalies are not covered. Hu et al [8] use spatiotemporal trajectory clustering to detect abnormality and predict paths. However, anomalies are only defined as low probable fittings. Piciarelli et al [11] proposed an online method to generate route models from key points. The order of apparition of the observations affects the model. Finally, a probabilistic method by Basharat et al [1] detects abnormality not only from trajectory positions but also from object sizes. However, this method is very dependent on a clean training dataset, classifying as anomaly slight variations of previously observed trajectories.

Next section describes the creation of scenario models from noisy trajectory training sets. Next, the hierarchy of anomalies is introduced. Subsequently, the performance of the proposed method is demonstrated. Last section concludes the paper and shows future lines of research.

Table 1 Comparison of previous approaches in terms of model, robustness and the types of anomalies detected.

Approach	Model used	Robust to noise	Types of anomaly
Fernyhough et al [6]	Vector quantization	No	HA
Makris et al [10]	GMM and vector quantization	Yes	HA
Hu et al [8]	Fuzzy C-means and GMM	Yes	SA, HA
Piciarelli et al [11]	Vector quantization	No	HA
Basharat et al [1]	Transition vectors and GMM	No	SA, HA
Our approach	GMM with Splines	Yes	SA, IA, HA

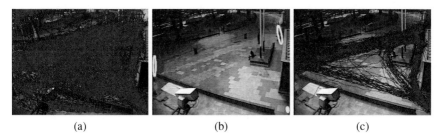

| (a) | (b) | (c) |

Fig. 1 (a) Original dataset of 4250 trajectories. (b) Detected Entry areas. (c) Dataset after removing noise caused by tracking failure and semistationary motion [10].

2 Scenario Model Learning

Our method learns from large recording datasets without previous manual selection of trajectories. Therefore, the initial dataset can contain spurious data, see Fig. 1(a). The most relevant noise issues are due to:

1. Failure of the motion-tracking algorithm. Errors may appear in form of *false* trajectories, where the motion history of multiple targets are mixed, or *split* trajectories that represent only a portion of the history.
2. Semistationary motion noise, e.g., trees, curtains, or window reflections. Apparent activity is detected in the vicinity of the noise source.
3. Non-smooth trajectories caused by inaccurate tracking or severe scenario conditions, causing irrealistic representations of motion in the scene.

The two first problems are solved by applying a multistep learning algorithm of Entry and Exit zones [10]. Fig. 1(c) shows the detected entry areas of the scenario. The sets S and E allow to obtain a subset $T' \subseteq T$ of trajectories that start in some entry of S and end in some exit of E, see Fig. 1.(d):

$$T' = \{t \in T \,|\, \exists_{s \in S, e \in E} : begin(t,s) \wedge end(t,e)\} \tag{1}$$

The third problem is solved by representing each T' with a continuous function model that solves tracking inaccuracies. We use a sequence of cubic splines [4], denoting as $s(t)$ the spline representation of trajectory t. The required number of cubic splines is automatically decided by computing the error between the original trajectory and the computed spline sequence:

$$error(t, s(t)) = \sum_{(x,y) \in t} d((x,y), s(t)) \tag{2}$$

Inner points are sampled with any required precision, maintaining the original temporal consistency. Moreover, it only requires storing intermediate control points and derivatives, thus reducing disk storage demand in large datasets.

2.1 Scenario Model

Next we detail the creation of route models from T'. A *route* $R_{s,e}$ between areas $(s,e) \in S \times E$ is defined using GMM as a *normal* way to go from s to e. Due to scene constraints or speed variations, there could be more than one route assigned to a pair (s,e). A *path* $P_{s,e}$ contains the routes from s to e. The final scene model includes all possible paths for each pair (s,e).

$$R_{s,e} = \{G_1, \ldots, G_k\} \tag{3}$$
$$P_{s,e} = \{R^1_{s,e}, \ldots, R^U_{s,e}\} \tag{4}$$
$$M = \{P_{s,e} | s \in S, e \in E\} \tag{5}$$

2.2 Learning Algorithm

The following procedure is applied to each pair $(s,e) \in S \times E$. Let $T_{s,e} = \{t_1, \ldots, t_N\} \subseteq T'$ be the set of trajectories starting at s and ending at e. Each trajectory t_n is represented by a sequence $r(t_n)$ of K equally spaced control points, obtained by sampling from $s(t_n)$, $r(t_n) = \{p^n_1, \ldots, p^n_K\}$, where p^n_k corresponds to the sample point k/K of $s(t_n)$.

The learning algorithm is sketched in 2(a): the input is a matrix of $K * N$ points, where each row k is the list of k-th control points of each trajectory in $T_{s,e}$. The path $P_{s,e}$ is initialized considering that all trajectories follow a single route. The algorithm traverses the lists of control points from 1 to K. At each step, the algorithm fits one- and two-component GMMs into the current point list, formed by the k-th control point of each trajectory, $\{p^1_k, \ldots, p^n_k\}$. The representativity of the model is evaluated through a density criterion:

$$\delta(G) = \frac{w}{\pi \cdot \sqrt{|\Sigma|}} \tag{6}$$

where w is the prior probability of G and Σ is its covariance matrix. If the mean density of the two-component GMM is higher than the single one, current route R^c splits into subroutes R^{c_1} and R^{c_2}, see Fig. 2(c). Once the algorithm has been applied to each pair (s,e), M contains the set of paths associated to *normality*. A trajectory deviating from M will be considered an anomaly. This result is enriched by considering the type of anomaly observed, as explained next.

Initialize $P_{s,e}$ with a single route R^1
for $k = 1$ to K **do**
   $nc \leftarrow |P_{s,e}|$
   for $c = 1$ to nc **do**
      $list \leftarrow points(k,c)$
      $G \leftarrow GMM(list,1)$
      $(G_1,G_2) \leftarrow GMM(list,2)$
      if $\delta(G) > (\delta(G_1) + \delta(G_2)) * \alpha/2$ **then**
         create R^{c_1}, R^{c_2} from R^c
         add G_1 to R^{c_1}
         add G_2 to R^{c_2}
         split $points(k,c)$ according to G_1 and G_2
         substitute R^c with R^{c_1}, R^{c_2} in $P_{s,e}$
      else
         add G to R^c
      end if
   end for
end for

(a)

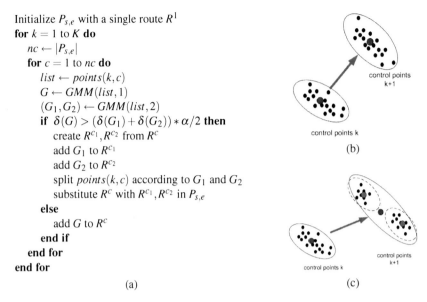

(b)

(c)

Fig. 2 (a) Route Modeling Algorithm for a pair $(s,e) \in S \times E$. It considers two possible situations: (b) the density of a single gaussian is enough to represent the k-th set of control points; or (c) two gaussians represent it better, so the route model splits.

3 Online Anomaly Detection

We distinguish three types of anomaly, see Fig. 3: (i) soft anomaly, **SA**: a modeled path is followed, i.e., the value of $p(P_{s,e}|\Theta,\omega)$ remains stable, but certain speed or orientation values differ from learned parameters; (ii) intermediate anomaly, **IA**: the most probable path $P_{s,e}$ changes during the development, for a whole window ω; and (iii) hard anomaly, **HA**: a completely unobserved path is followed, which can be caused by $e' \notin E$ or because the probability of having started from s is too low for the whole window ω.

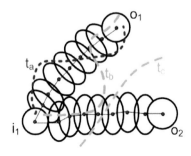

Fig. 3 Types of anomalies: t_a incurs in *SA* by partially falling out of a route. t_b shows *IA* because of changing paths. t_c deviates from all routes, thus producing *HA*.

This hierarchy provides degrees of deviations between new observations and the model. Indeed, a SA can be considered as a route deviation inside a path; a IA detects path deviations within M; and a HA complete deviates from M. The prior probability that a pixel location (x, y) is a part of a modeled path, $p(P_{s,e}|(x, y))$, is obtained by computing a probabilistic map of the image plane. These probabilities are stored in a $h \times w \times |M|$ matrix Θ, where h, w are the dimensions of the image and $|M|$ is the number of paths:

$$\Theta_{x,y} = p(P_{s,e}|(x,y)) = max\,(p(G_j|(x,y)) \tag{7}$$

where G_j belongs to some route $R_{s,e}^u \in P_{s,e}$.

Anomaly detection is performed online as new trajectories are available. Given the current observation, we maintain a temporal window ω with the last $|\omega|$ observations. We compute the probability of being in a given path by assuming correlation with previous frame steps:

$$p(P_{s,e}|\Theta, \omega, (x,y)) = \alpha * \Theta_{x,y} + (1 - \alpha) * \frac{\sum_i^{|w|} p(P_{s,e}|\Theta, w_i)}{|w|} \tag{8}$$

where w_i is the i–th observation in ω, and the update factor α is set to 0.5.

The anomaly detection algorithm labels trajectories as new tracking observations arrive. A new trajectory t is associated to one of the learned entries $s \in S$. If no entry is close enough, t is labeled as HA. Otherwise, the probabilities of being in each path containing s are computed using Eq. 8, for each new observation (x, y). In case of similar probabilities, t is not assumed to follow any concrete path.

When the probability of path $P_{s,e}$ is higher than the others, the trajectory is assumed to be following it for the next frames. If a different path becomes more probable t starts being labeled as IA until the initial path becomes the most probable again; if t finally ends in e', then the IA period is that in which $P_{s,e}$ was less probable. Finally, if all paths are improbable for a given ω, t is marked as HA until a probable path is assigned. Assuming a path $P_{s,e}$, we select the gaussian G^* that better represents the current observation:

$$j^* = \arg\max_{j}\ p(G_j|(x,y)) \tag{9}$$

To know if path $P_{s,e}$ is being followed normally, we compute the sum of probability increments that ω is between Gaussians G_{j^*-1} and G_{j^*+1}, previous and next on the path. This is encoded into a *normality* factor $F(\omega)$:

$$F(\omega) = \begin{cases} \sum_i^{|\omega|} -\Delta p(G_{j^*-1}|\omega_i) & \text{if } h(G_{j^*}, \omega) > 0 \\ \sum_i^{|\omega|} \Delta p(G_{j^*+1}|\omega_i)) & \text{otherwise} \end{cases} \tag{10}$$

where $\Delta p(G|\omega_i) = p(G|\omega_i) - p(G|\omega_{i-1})$, and $h(G_{j^*}, \omega) = p(G_{j^*}|\omega_{|\omega|}) - p(G_{j^*}|\omega_1)$ measures the direction of ω towards G_{j^*}. Positive values of $F(\omega)$ certify that the last ω observations are normal according to M. Otherwise, the trajectory deviates from

M, so we annotate subsequent frames as SA until the deviation is over. Note that $F(\omega)$ is only representative when a path is followed, i.e., the path has been the most probable during ω. Thus, when labeling as IA or a HA, the value of $F(\omega)$ does not provide any extra information.

4 Results

We evaluate the method using a real database from [2]. The dataset was cleaned as detailed in the text, to avoid the three sources of noise described. A wide range of values were tested for K –number of control points per path– and ω –length of

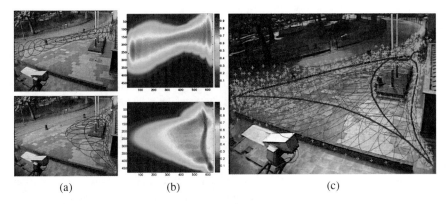

(a)   (b)   (c)

Fig. 4 (a) Learned routes, (b) their Θ maps, and (c) the final model.

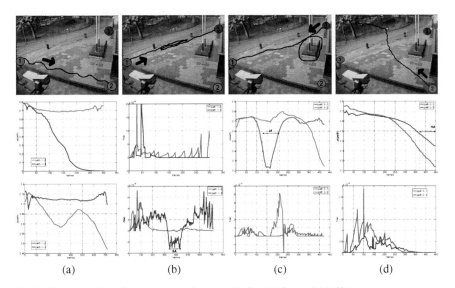

(a)   (b)   (c)   (d)

Fig. 5 Top: examples of (a) normal trajectory, (b) SA, (c) IA, and (d) HA.

temporal window–. Here we use $K = 10$ and $|\omega| = 20$. Fig. 4 shows results for two paths, their probabilistic maps Θ, and the resulting model M.

Path P_1 is more probable than P_2 if $p(P_1|\Theta, \omega, (x,y)) > 1.5 * p(P_1|\Theta, \omega, (x,y))$. A trajectory is considered HA if all paths have probability below 0.5. Fig. 5 shows four results: the first row are trajectories in the image plane. The second and third rows show the temporal evolution of $p(P_{s,e}|\Theta, \omega, (x,y))$ and $F(\omega)$, respectively. Case (a) represents a normal trajectory –single probable path, $F(\omega) > 0$–. Case (b) also has a single assignment, but $F(\omega) < 0$ when the target goes backwards –interval marked as SA–. Case (c) shows an IA, since the assignment changes and stabilizes; it is unclear which part is the anomalous one. Finally, case (d) is a HA, since the trajectory differs from any path having entry (2).

5 Conclusions

We provided a trajectory-based definition of anomaly in video surveillance. Our hierarchy classifies anomalies regarding the extent and nature of the deviations from a learned model. A novel GMM representation of observed common routes creates probabilistic maps that typify abnormality in real-time. The characteristics of our method overcomes limitations of similar existing approaches, and performs well in spite of noisy tracking conditions. This technique has direct applications in the area of intelligent video surveillance, by providing a richer semantic explanation of observed abnormality than existing approaches. However, there is still a gap between the notion of anomaly in terms of attentional interest and its statistical definition; future work will refine the proposal with a conceptual layer exploiting common knowledge about suspicious behaviors.

Acknowledgments. This work was originally supported by EC grant IST-027110 HERMES project; and by the Spanish projects Avanza I+D ViCoMo (TSI-020400-2009-133), TIN2009-14501-C02 and CONSOLIDER-INGENIO 2010 MIPRCV (CSD2007-00018).

References

1. Basharat, A., Gritai, A., Shah, M.: Learning object motion patterns for anomaly detection and improved object detection. In: CVPR, Anchorage, USA (2008)
2. Black, J., Makris, D., Ellis, T.: Hierarchical database for a multi-camera surveillance system. Pattern Analysis and Applications 7(4), 430–446 (2004)
3. Bremond, F., Thonnat, M., Zuniga, M.: Video understanding framework for automatic behavior recognition. Behavior Research Methods 3(38), 416–426 (2006)
4. de Boor, C.: A practical guide to splines. Springer, New York (1978)
5. Fernández, C., Baiget, P., Roca, F.X., Gonzàlez, J.: Interpretation of Complex Situations in a Semantic-Based Surveillance Framework. Signal Processing: Image Communication 23(7), 554–569 (2008)
6. Fernyhough, J.H., Cohn, A.G., Hogg, D.: Generation of semantic regions from image sequences. In: Buxton, B.F., Cipolla, R. (eds.) ECCV 1996. LNCS, vol. 1065, pp. 475–484. Springer, Heidelberg (1996)

7. Gonzàlez, J., Rowe, D., Varona, J., Roca, F.: Understanding dynamic scenes based on human sequence evaluation. IMAVIS 27(10), 1433–1444 (2009)
8. Hu, W., Xiao, X., Fu, Z., Xie, D.: A system for learning statistical motion patterns. IEEE TPAMI 28(9), 1450–1464 (2006)
9. Johnson, N., Hogg, D.: Learning the distribution of object trajectories for event recognition. In: BMVC 1995, pp. 583–592. BMVA Press, Surrey (1995)
10. Makris, D., Ellis, T.: Learning semantic scene models from observing activity in visual surveillance. IEEE TSMC–Part B 35(3), 397–408 (2005)
11. Piciarelli, C., Foresti, G.L.: On-line trajectory clustering for anomalous events detection. PRL 27(15), 1835–1842 (2006)

Identifying Surprising Events in Video Using Bayesian Topic Models*

Avishai Hendel, Daphna Weinshall, and Shmuel Peleg

Abstract. In this paper we focus on the problem of identifying interesting parts of the video. To this end we employ the notion of Bayesian surprise, as defined in [9, 10], in which an event is considered surprising if its occurrence leads to a large change in the probability of the world model. We propose to compute this abstract measure of surprise by first modeling a corpus of video events using the Latent Dirichlet Allocation model. Subsequently, we measure the change in the Dirichlet prior of the LDA model as a result of each video event's occurrence. This leads to a closed form expression for an event's level of surprise. We tested our algorithm on a real world video data, taken by a camera observing an urban street intersection. The results demonstrate our ability to detect atypical events, such as a car making a U-turn or a person crossing an intersection diagonally.

1 Introduction

1.1 Motivation

The availability and ubiquity of video from security and monitoring cameras has increased the need for automatic analysis and classification. One urging problem is that the sheer volume of data renders it impossible for human viewers, the ultimate classifiers, to watch and understand all of the displayed content. Consider for example a security officer who may need to browse through the hundreds of cameras positioned in an airport, looking for possible suspicious activities - a laborious task that is error prone, yet may be life critical. In this paper we address the problem of unsupervised video analysis, having applications in various domains, such as the

Avishai Hendel · Daphna Weinshall · Shmuel Peleg
School of Computer Science and Engineering, Hebrew University of Jerusalem, Israel
e-mail: avishai.hendel@mail.huji.ac.il, daphna@cs.huji.ac.il,
    peleg@mail.huji.ac.il

* Work is funded by the EU Integrated Project DIRAC (IST-027787).

D. Weinshall, J. Anemüller, and L. van Gool (Eds.): DIRAC, SCI 384, pp. 97–105.
springerlink.com

inspection of surveillance videos, examination of 3D medical images, or cataloging and indexing of video libraries.

A common approach to video analysis serves to assist human viewers by making video more accessible to sensible inspection. In this approach the human judgment is maintained, and video analysis is used only to assist viewing. Algorithms have been devised to create a compact version of the video, where only certain activities are displayed [1], or where all activities are displayed using video summarization [2].

We would like to go beyond summarization; starting from raw video input, we seek an automated process that will identify the unusual events in the video, and reduce the load on the human viewer. This process must first extract and analyze activities in the video, followed by establishing a model that characterizes these activities in a manner that permits meaningful inference. A measure to quantify the significance of each activity is needed as a last step.

1.2 Our Approach

We present a generative probabilistic model that accomplishes the tasks outlined above in an unsupervised manner, and test it in a real world setting of a webcam viewing an intersection of city streets.

The preprocessing stage consists of the extraction of video activities of high level objects (such as vehicles and pedestrians) from the long video streams given as input. Specifically, we identify a set of video events (video tubes) in each video sequence, and represent each event with a 'bag of words' model. We introduce the concept of 'transition words', which allows for a compact, discrete representation of the dynamics of an object in a video sequence. Despite its simplicity, this representation is successful in capturing the essence of the input paths. The detected activities are then represented using a latent topic model, a paradigm that has already shown promising results [13, 8, 7, 20].

Next, we examine the video events in a rigorous Bayesian framework, to identify the most interesting events present in the input video. Thus, in order to differentiate intriguing events from the typical commonplace events, we measure the effect of each event on the observer's beliefs about the world, following the approach put forth in [9, 10]. We propose to measure this effect by comparing the prior and posterior parameters of the latent topic model, which is used to represent the overall data. We then show that in the street camera scenario, our model is able to pick out atypical activities, such as vehicle U-turns or people walking in prohibited areas.

2 Activity Representation

2.1 Objects as Space Time Tubes

The fundamental representation of objects in our model is that of 'video tubes' [3]. A tube is defined by a sequence of object masks carved through the space time volume, assumed to contain a single object of interest (e.g., in the context of street cameras, it may be a vehicle or a pedestrian). This localizes events in both space and time, and enables the association of local visual features with a specific object, rather than an entire video.

A modification of the 'Background Cut' method [4] is used to distinguish foreground blobs from the background. The blobs are then matched by spatial proximity to create video tubes that extend through time.

2.2 Trajectories

An obvious and important characteristic of a video tube is its trajectory, as defined by the sequence of its spatial centroids. A preferable encoding in our setting should capture the characteristic of the tube's path in a compact and effective way, while considering location, speed and form.

Of the numerous existing approaches, we use a modification of the method suggested in [5]. The process is summarized in Fig. 1. The displacement vector of the object's centroid between consecutive frames is quantized into one of 25 bins, including a bin for zero displacement. A transition occurrence matrix, indicating the frequency of bin transitions in the tube is assembled, and regarded as a histogram of 'transition words', where each word describes the transition between two consecutive quantized displacement vectors. The final representation of a trajectory is this histogram, indicating the relative frequency of the 625 possible transitions.

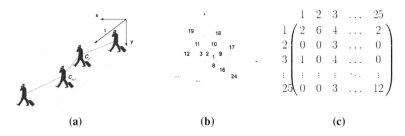

(a)　　　　　　　　　(b)　　　　　　　　　(c)

Fig. 1 Trajectory representation: the three stages of our trajectory representation: (a) compute the displacement of the centroids of the tracked object between frames, (b) quantize each displacement vector into one of 25 quantization bins, and (c) count the number of different quantization bin transitions in the trajectory into a histogram of bin transitions.

3 Modeling of Typical Activities Using LDA

We use the *Latent Dirichlet Allocation* (LDA) model as our basis for the representation of the environment and events present in the input video. The model, which was first introduced in the domain of text analysis and classificiation [6] has been successfully applied recently to computer vision tasks, where the text topics have been substituted by scenery topics [7] or human action topics [8]. As noted above, each tube is represented as a histogram of transition words taken from the trajectory vocabulary $V = \{w_{1-1}, w_{1-2}, ..., w_{25-24}, w_{25-25}\}, |V| = 625$. A set of video tubes $T = \{T_1, T_2, ..., T_m\}$ is given as input to the standard LDA learning procedure, to obtain the model's parameters α and β. These parameters complete our observer's model of the world.

The Dirichlet prior α describes the common topic mixtures that are to be expected in video sequences taken from the same source as the training corpus. A specific topic mixture θ_t determines the existence of transitions found in the trajectory using the per-topic word distribution matrix β. The actual mixture of an input tube is intractable, but can be approximated by an optimization problem that yields the posterior Dirichlet parameter γ_t^*.

4   Surprise Detection

The notion of surprise is, of course, human-centric and not well defined. Surprising events are recognized as such with regard to the domain in question, and background assumptions that can not always be made explicit. Thus, rule based methods that require manual tuning may succeed in a specific setting, but are doomed to failure in less restricted settings. Statistical methods, on the other hand, require no supervision. Instead, they attempt to identify the expected events from the data itself, and use this automatically learned notion of typicality to recognize the extraordinary events.

Such framework is proposed in the work by Itti [9] and Schmidhuber [10]. Dubbed 'Bayesian Surprise', the main conjecture is that a surprising event from the viewpoint of an observer is an event that modifies its current set of beliefs about the environment in a significant manner. Formally, assume an observer has a model M to represent its world. The observer's belief in this model is described by the prior probability of the model $p(M)$ with regard to the entire model space $\mathcal{M}$. Upon observing a new measurement t, the observer's model changes according to Bayes' Law:

$$p(M \mid t) = \frac{p(M)p(t \mid M)}{p(t)} \tag{1}$$

This change in the observer's belief in its current model of the world is defined as the surprise experienced by the observer. Measurements that induce no or minute changes are not surprising, and may be regarded as 'boring' or 'obvious' from the observer's point of view. To quantify this change, we may use the KL divergence between the prior and posterior distributions over the set $\mathcal{M}$ of all models:

$$S(t, M) = KL(p(M), p(M \mid t)) = \int_{\mathcal{M}} p(M) log \frac{p(M)}{p(M \mid t)} dM \tag{2}$$

This definition is intuitive in that surprising events that occur repeatedly will cease to be surprising, as the model is evolving. The average taken over the model space also ensures that events with very low probability will be regarded as surprising only if they induce a meaningful change in the observer's beliefs, thus ignoring noisy incoherent data that may be introduced.

Although the integral in Eq. (2) is over the entire model space, turning this space to a parameter space by assuming a specific family of distributions may allow us to compute the surprise measure analytically. Such is the case with the Dirichlet family of distributions, which has several well known computational advantages: it is in the exponential family, has finite dimensional sufficient statistics, and is conjugate to the multinomial distribution.

5 Bayesian Surprise and the LDA Model

As noted above, the LDA model is ultimately represented by its Dirichlet prior α over topic mixtures. It is a natural extension now to apply the Bayesian surprise framework to domains that are captured by LDA models.

Recall that video tubes in our 'bag of words' model are represented by the posterior optimizing parameter γ^*. Furthermore, new evidence also elicits a new Dirichlet parameter for the world model of the observer, $\widehat{\alpha}$. To obtain $\widehat{\alpha}$, we can simulate one iteration of the variational EM procedure used above in the model's parameters estimation stage, where the word distribution matrix β is kept fixed. This is the Dirichlet prior that would have been calculated had the new tube been appended to the training corpus. The Bayesian Surprise formula when applied to the LDA model can be now written as:

$$S(\alpha, \widehat{\alpha}) = KL_{DIR}(\alpha, \widehat{\alpha}) \tag{3}$$

The Kullback - Leibler divergence of two Dirichlet distributions can be computed as [11]:

$$KL_{DIR}(\alpha, \widehat{\alpha}) = log\frac{\Gamma(\alpha)}{\Gamma(\widehat{\alpha})} + \sum_{i=1}^{k} log\frac{\Gamma(\widehat{\alpha}_i)}{\Gamma(\alpha_i)} + \sum_{i=1}^{k}[\alpha_i - \widehat{\alpha}_i][\psi(\alpha_i) - \psi(\alpha)] \tag{4}$$

where

$$\alpha = \sum_{i=1}^{k} \alpha_i \quad \text{and} \quad \widehat{\alpha} = \sum_{i=1}^{k} \widehat{\alpha}_i$$

and Γ and ψ are the gamma and digamma functions, respectively.

Thus each video event is assigned a surprise score, which reflects the tube's deviation from the expected topic mixture. In our setting, this deviation may correspond to an unusual trajectory taken by an object, such as 'car doing a U-turn', or 'person running across the road'. To obtain the most surprising events out of a corpus, we can select those tubes that receive a surprise score that is higher than some threshold.

6 Experimental Results

6.1 Dataset

We test our model on data obtained from a real world street camera, overlooking an urban road intersection. This scenario usually exhibits structured events, where pedestrians and vehicles travel and interact in mostly predefined ways, constrained by the road and sidewalk layout. Aside from security measures, intersection monitoring has been investigated and shown to help in reducing pedestrian and vehicle conflicts, which may result in injuries and crashes [18].

6.2 Trajectory Classification

The first step in our algorithm is the construction of a model that recognizes typical trajectories in the input video. We fix k, the number of latent topics to be 8. Fig. 2 shows several examples of classified objects from four of the eight model topics, including examples from both the training and test corpora.

Note that some of the topics seem to have a semantic meaning. Thus, on the basis of trajectory description alone, our model was able to automatically catalog the video tubes into semantic movement categories such as 'left to right', or 'top to bottom', with further distinction between smooth constant motion (normally cars) and the more erratic path typically exhibited by people. It should be noted, however, that not all latent topics correspond with easily interpretable patterns of motion as depicted in Fig. 2. Other topics seem to capture complicated path forms, where pauses and direction changes occur, with one topic representing 'standing in place' trajectories.

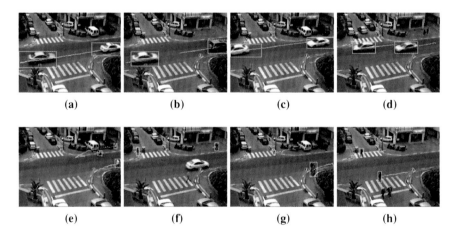

Fig. 2 Trajectory classifications: (a,b) cars going left to right, (c,d) cars going right to left, (e,f) people walking left to right, and (g,h) people walking right to left.

6.3   Surprising Events

To identify the atypical events in the corpus, we look at those tubes which have the highest surprise score. Several example tubes which fall above the 95th percentile are shown in Fig. 4. They include such activities as a vehicle performing a U-turn,

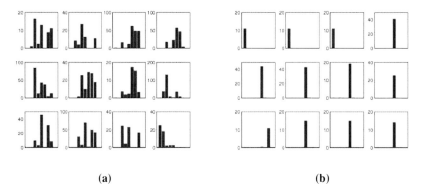

(a)   (b)

Fig. 3 Posterior Dirichlet parameters γ^* values for the most surprising (a) and typical (b) events. Each plot shows the values of each of the $k = 8$ latent topics. Note that the different y scales correspond to different trajectory lengths (measured in frames).

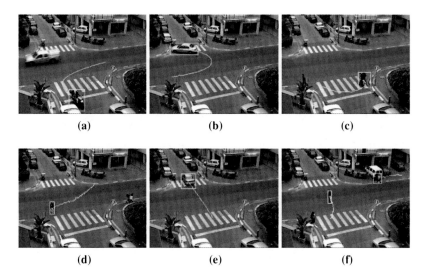

Fig. 4 Surprising events: (a) a bike turning into a one-way street from the wrong way, (b) a car performing a U-turn, (c) a bike turning and stalling over pedestrian crossing, (d) a man walking across the road, (e) a car crossing the road from bottom to top, (f) a woman moving from the sidewalk to the middle of the intersection.

or a person walking in a path that is rare in the training corpus, like crossing the intersection diagonally.

In Fig. 3 the γ^* values of the most surprising and typical trajectories are shown. It may be noted that while 'boring' events generally fall into one of the learned latent topics exclusively (Fig. 3b), the topic mixture of surprising events has massive counts in several topics at once (Fig. 3a). This observation is verified by computing the mean entropy measure of the γ^* parameters, after being normalized to a valid probability distribution:

$$\overline{H}(\gamma_{surprising}) = 1.2334, \quad \overline{H}(\gamma_{typical}) = 0.5630$$

7 Conclusions

In this work we presented a novel integration between the generative probabilistic model LDA and the Bayesian surprise framework. We applied this model to real world data of urban scenery, where vehicles and people interact in natural ways. Our model succeeded in automatically obtaining a concept of the normal behaviors expected in the tested environment, and in applying these concepts in a Bayesian manner to recognize those events that are out of the ordinary. Although the features used are fairly simple (the trajectory taken by the object), complex surprising events such as a car stalling in its lane, or backing out of its parking space were correctly identified, judged against the normal paths present in the input.

References

1. Boiman, O., Irani, M.: Detecting Irregularities in Images and in Video. International Journal of Computer Vision 74(1), 17–31 (2007)
2. Pritch, Y., Rav-Acha, A., Peleg, S.: Nonchronological Video Synopsis and Indexing. IEEE Trans. Pattern Anal. Mach. Intell. 30(11), 1971–1984 (2008)
3. Pritch, Y., Ratovitch, S., Hendel, A., Peleg, S.: Clustered Synopsis of Surveillance Video. In: Proc. AVSS (2009)
4. Sun, J., Zhang, W., Tang, X., Shum, H.-Y.: Background cut. In: Leonardis, A., Bischof, H., Pinz, A. (eds.) ECCV 2006. LNCS, vol. 3952, pp. 628–641. Springer, Heidelberg (2006)
5. Sun, J., Wu, X., Yan, S., Cheong, L.F., Chua, T.S., Li, J.: Hierarchical spatio-temporal context modeling for action recognition. In: Proc. CVPR (2009)
6. Blei, D.M., Ng, A.Y., Jordan, M.I.: Latent Dirichlet Allocation. In: Proc. NIPS (2001)
7. Fei-Fei, L., Perona, P.: A Bayesian Hierarchical Model for Learning Natural Scene Categories. In: Proc. CVPR (2005)
8. Niebles, J.C., Wang, H., Fei-Fei, L.: Unsupervised Learning of Human Action Categories Using Spatial-Temporal Words. International Journal of Computer Vision 79(3), 299–318 (2008)
9. Itti, L., Baldi, P.: A Principled Approach to Detecting Surprising Events in Video. In: Proc. CVPR (2005)

10. Schmidhuber, J.: Driven by Compression Progress: A Simple Principle Explains Essential Aspects of Subjective Beauty, Novelty, Surprise, Interestingness, Attention, Curiosity, Creativity, Art, Science, Music, Jokes. In: Proc. ABiALS (2008)
11. Penny, W.D.: Kullback-Liebler Divergences of Normal, Gamma, Dirichlet and Wishart Densities. Technical report, Wellcome Department of Cognitive Neurology (2001)
12. Blei, D.M., Griffiths, T.L., Jordan, M.I.: The nested chinese restaurant process and bayesian nonparametric inference of topic hierarchies. J. ACM 57(2) (2010)
13. Sivic, J., Russell, B.C., Efros, A.A., Zisserman, A., Freeman, W.T.: Discovering Objects and their Localization in Images. In: Proc. ICCV (2005)
14. Lowe, D.G.: Distinctive Image Features from Scale-Invariant Keypoints. International Journal of Computer Vision 60(2), 91–110 (2004)
15. Laptev, I., Lindeberg, T.: Local Descriptors for Spatio-temporal Recognition. In: Proc. SCVMA (2004)
16. Hofmann, T.: Probabilistic Latent Semantic Analysis. In: Proc. UAI (1999)
17. Jordan, M.I., Ghahramani, Z., Jaakkola, T., Saul, L.K.: An Introduction to Variational Methods for Graphical Models. Machine Learning 37(2), 183–233 (1999)
18. Hughes, R., Huang, H., Zegeer, C., Cynecki, M.: Evaluation of Automated Pedestrian Detection at Signalized Intersections. United States Department of Transportation, Federal Highway Administration Report (2001)
19. Ranganathan, A., Dellaert, F.: Bayesian surprise and landmark detection. In: Proc. ICRA (2009)
20. Hospedales, T., Gong, S., Xiang, T.: A Markov Clustering Topic Model for Mining Behaviour in Video. In: Proc. ICCV (2009)
21. Wang, X., Ma, X., Grimson, E.: Unsupervised Activity Perception by Hierarchical Bayesian Models. In: Proc. CVPR (2007)

Part IV
Dealing with Meaningful Novel Events, What to Do after Detection

Anomaly Detection and Knowledge Transfer in Automatic Sports Video Annotation

I. Almajai, F. Yan, T. de Campos, A. Khan, W. Christmas,
D. Windridge, and J. Kittler

Abstract. A key question in machine perception is how to adaptively build upon existing capabilities so as to permit novel functionalities. Implicit in this are the notions of *anomaly detection* and *learning transfer*. A perceptual system must firstly determine at what point the existing learned model ceases to apply, and secondly, what aspects of the existing model can be brought to bear on the newly-defined learning domain. *Anomalies* must thus be distinguished from mere *outliers*, i.e. cases in which the learned model has failed to produce a clear response; it is also necessary to distinguish novel (but meaningful) input from misclassification error within the existing models. We thus apply a methodology of anomaly detection based on comparing the outputs of strong and weak classifiers [10] to the problem of detecting the rule-incongruence involved in the transition from singles to doubles tennis videos. We then demonstrate how the detected anomalies can be used to transfer learning from one (initially known) rule-governed structure to another. Our ultimate aim, building on existing annotation technology, is to construct an adaptive system for court-based sport video annotation.

1 Introduction

Artificial cognitive systems should be able to autonomously extend capabilities to accommodate anomalous input as a matter of course (humans are known to be able to establish novel categories from single instances [8]). The anomaly detection problem is typically one of distinguishing novel (but meaningful) input from misclassification error within existing models. By extension, the *treatment* of anomalies so determined involves adapting the existing domain model to accommodate the anomalies in a robust manner, maximising the transfer of learning from the original domain so as to avoid over-adaptation to outliers (as opposed to merely incongruent events). That is, we seek to make conservative assumptions when adapting the system.

I. Almajai · F. Yan · T. de Campos · A. Khan · W. Christmas · D. Windridge · J. Kittler
CVSSP, University of Surrey, Guildford GU2 7XH, UK
e-mail: d.windridge@surrey.ac.uk

D. Weinshall, J. Anemüller, and L. van Gool (Eds.): DIRAC, SCI 384, pp. 109–117.

The composite system for detecting and treating anomaly should thus be capable of bootstrapping novel representations via the interaction between the two processes. In this paper, we aim to demonstrate this principle with respect to the redefinition of key entities designated by the domain rules, such that the redefinition renders the existing rule base *non-anomalous*. We thus implicitly designate a new domain (or context) by the application of anomaly detection.

Our chosen framework for anomaly detection is that advocated in [10, 15] which distinguishes outliers from anomalies via the disparity between a generalised context classifier (when giving a low confidence output) and a combination of 'specific-level' classifiers (generating a high confidence output). The classifier disparity leading to the anomaly detection can equally be characterised as being between strongly constrained (contextual) and weakly constrained (non-contextual) classifiers [2]. A similar approach can be used for model updating and acquisition within the context of tracking [13] and for the simultaneous learning of motion and appearance [14]. Such tracking systems explicitly address the *loss-of-lock* problem that occurs without model updating.

In this paper we consider anomaly detection in the context of sporting events. What we propose here is a system that will detect when the rules of tennis matches change. We start with a system trained to follow singles matches, and then change the input material to doubles matches. The system should then start to flag anomalies, in particular, events relating to the court area considered to be "in play".

The system is based on an existing tennis annotation system [5, 11], which is used to generate data for the anomaly detection. This system provides basic video analysis tools: de-interlacing, lens correction, and shot segmentation and classification. It computes a background mosaic, which it uses to locate foreground objects and hence track the players. By locating the court lines, it computes the projection between the camera and ground plane using a court model. It is also able to track the ball; this is described in more detail in the next section.

In the next section we describe the weak classifiers and their integration. In Section 3 we discuss the problem anomaly detection mechanism. We describe some experiments to validate the ideas in Section 4, incorporating the results into the anomaly-adaptation/rule-update stage in the immediately following section. We conclude the paper in Section 6.

2   Weak Classifiers

2.1   *Ball Event Recognition*

Ball event recognition is one of the weak classifiers we employ. In the following, we first briefly describe the tennis ball tracker, then introduce an HMM-based ball event classifier.

Tennis ball tracking: To detect the key ball events that describe how the match progresses, e.g. the tennis ball being hit or bouncing on the ground, the tracking of the tennis ball in the play shots is required. This is a challenging task: small objects usually have fewer features to detect and are more vulnerable to distractions; the movement of the tennis ball is so fast that sometimes it is blurred into the background, and is also subject to temporary occlusion and sudden change of motion direction. Even worse, motion blur, occlusion, and abrupt motion change tend to happen together: when the ball is close to one of the players. To tackle these difficulties, we propose a ball tracker based on [11] with the following sequence of operations: (i) Candidate blobs are found by background subtraction. (ii) Blobs are then classified as ball / not ball using their size, shape and gradient direction at blob boundary. (iii) "Tracklets" are established in the form of 2nd-order (i.e. roughly parabolic) trajectories. These correspond to intervals when the ball is in free flight. (iv) A graph-theoretic data association technique is used to link tracklets into complete ball tracks. Where the ball disappears off the top of the frame and reappears, the tracks are linked. (v) By analysing the ball tracks, sudden changes in velocity are detected as "ball events". These events will be classified in an HMM-based classifier, to provide information of how the tennis game progresses.

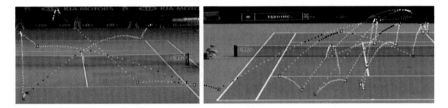

Fig. 1 Two examples of the final ball tracking results with ball event detection. Yellow dots: detected ball positions. Black dots: interpolated ball positions. Red squares: detected ball events. In the left example, there is one false positive and one false negative in ball event detection. In the right example, there are a few false negatives.

HMM-based ball event recognition: The key event candidates of the ball tracking module need to be classified into serve, bounce, hit, net, etc. The higher the accuracy of the event detection and classification stage the less likely that the high level interpretation module may misinterpret the event sequences. A set of continuous-density left-to-right first-order HMMs, $\Lambda = \{\lambda_1, \ldots, \lambda_k, \ldots, \lambda_K\}$ are used to analyse the ball trajectory dynamics and recognise events regionally within the tracked ball trajectory, based on [1], but using the detected ball motion changes to localise events. K is the number of event types in a tennis game, including a null event needed to identify false positives in event candidates. An observation, $\mathbf{o}_t$, at time t, is composed of the velocity and acceleration of the ball position in the mosaic domain: $\mathbf{o}_t = \{\dot{\mathbf{x}}_t, \ddot{\mathbf{x}}_t\}$. To classify an event at a time t, a number of observations $\mathbf{O}_t = \mathbf{o}_{t-W}, \mathbf{o}_{t-W+1}, \ldots, \mathbf{o}_{t+W}$ are considered within a window of size $2W + 1$. Each HMM is characterised by three probability measures: the state transition probability distribution matrix A, the observation probability distribution B and the initial state distribution π, defined for a set of N states $S = (s_1, s_2, \ldots, s_N)$, and ball information observation sequence $\mathbf{O}_t$. Each state s_j is represented by a number, M_j, of Gaussian mixture components. Given a

set of training examples corresponding to a particular model, the model parameters
are determined by the Baum-Welch algorithm [12]. Thus, provided that a sufficient
number of representative examples of each event can be collected, an HMM can be
constructed which implicitly models the sources of variability in the ball trajectory
dynamics around events. Once the HMMs are trained, the most likely state sequence
for a new observation sequence is calculated for each model using the Viterbi algo-
rithm [12]. The event is then classified by computing $\hat{k} = \arg\max_k (P(\mathbf{O}_t|\lambda_k))$.

For every ball trajectory, the first task is to identify when the serve takes place.
The first few key event candidates are searched; once the serve position and time
are determined, the subsequent key events candidates are classified into their most
probable categories, and the null events, considered false positives, are ignored. For
recognised bounce events, the position of the ball bounce on the court is determined
in court coordinates. This process can have some false negatives due to ball occlusion and smooth interpolation etc. [11]. This happens often when there is a long time
gap between two recognised events. To recover from such suspected false negatives,
an exhaustive search is used to find other likely events in such gaps.

2.2 Action Recognition

In tennis games the background can easily be tracked, which enables the use of
heuristics to robustly segment player candidates, as explained in [5]. To reduce the
number false positives, we extract bounding boxes of the moving blobs and merge
the ones that are close to each other. Next, geometric and motion constraints are
applied to further remove false positives. A map of likely player positions is built
by cumulating player bounding boxes from the training set. A low threshold on this
map disregards bounding boxes from the umpires and ball boys/girls. In subsequent
frames, the players are tracked with a particle filter. Fig. 2 shows some resulting
players detected in this manner, performing different actions.

Fig. 2 Sample images and detected players performing each primitive action of tennis.

Given the location of each player, we extract a single spatio-temporal descriptor
at the centre of the player's bounding box, with a spatial support equal to the maximum between the width and height of the box. The temporal support was set to 12
frames. This value was determined using the validation set of the KTH dataset. As a
spatio-temporal descriptor, we chose the 3DHOG (histogram of oriented gradients)
method of Klaser et al. [6]. This method gave state-of-the-art results in recent bench-
marks of Wang et al. [9] and has a number of advantages in terms of efficiency and

stability over other methods. Previously, 3DHOG has only been evaluated in bag-of-visual-words (BoW) frameworks. In [3], we observed that if players are detected, a single 3DHOG feature extraction followed by classification with kernel LDA gives better performance than an approach based on BoW with keypoint detection.

Three actions are classified: *serve*, *hit* and *non-hit*. A *hit* is defined by the moment a player hits the ball, if this is not a *serve* action. *Non-hit* refers to any other action, e.g. if a player swings the racket without touching the ball. Separate classifiers are trained for near and far players. Their location w.r.t. the court lines is easily computed given the estimated projection matrix. For training, we used only samples extracted when a change of ball velocity is detected. For the test sequences, we output results for every frame. Classification was done with Kernel LDA using a one-against-rest set-up.

We determine the classification results using a majority voting scheme in a temporal window. We also post-process them by imposing these constraints that are appropriate for court games: (i) players are only considered for action classification if they are close to the ball, otherwise the action is set to *non-hit*; (ii) at the beginning of a play shot, we assume that detected *hits* are actually *serves*; (iii) at later moments, serves are no longer enabled, i.e., if a serve is detected later in a play shot, the action is classified as *hit*. This enables overhead-hits (which are visually the same as *serves*) to be classified as *hits*.

In order to provide a confidence measure for the next steps of this work, we use the classification scores from KLDA (normalised distance to the decision boundary).

2.3   Bounce Position Uncertainty

As the ball position measurements and camera calibration are subject to errors, the probability values near the boundaries of parts of the court will bleed into the neighbouring regions. This can be modelled by a convolution between the probability function $\hat{P}(bounce_{in}|\mathbf{x}_t)$, where $\mathbf{x}_t$ is the ball position, and the measurement error function $p(e)$ which is assumed to be Gaussian with zero mean and standard deviation σ_{ball}.

$$P(bounce_{in}|\mathbf{x}_t) = \int_{\psi} \hat{P}(bounce_{in}|\psi)p(\mathbf{x}_t - \psi)d\psi \tag{1}$$

Finally, the probability of $bounce_{out}$ is given by $P(bounce_{out}|\mathbf{x}_t) = 1 - P(bounce_{in}|\mathbf{x}_t)$.

2.4   Combining Evidence

The sequence of events is determined by the output of the ball event recognition HMM. Firstly a serve is searched for. If "serve" is one of the 4 most probable HMM hypotheses of an event, that event is deemed to be the serve, and the search is terminated. The remaining events are then classified on the basis of the most probable HMM hypothesis. The entire ball trajectory is then searched for possible missed events: e.g. if consecutive events are bounces on opposite ends of the court, it is likely that a hit was missed.

Sequences of events that start with a serve are passed to the context classification stage. Event sequences are composed of some 17 event types (see [7]). Each event is assigned a confidence, based on the HMM posterior probabilities and, for hits, the action confidences. The combination rules are at present a set of Boolean heuristics, based on human experience. Bounce events are also assigned a separate confidence, based on the bounce position uncertainty.

3   Context Classification

To detect incongruence, we devise an HMM stage similar to the high-level HMM used to follow the evolution of a tennis match as described in [7]. Here, each sequence of events starting with a serve is analysed to see if it is a failed serve or a point given to one of the two opponents. The aim is to find sequences of events in which the temporal context classifier reaches a decision about awarding a point before the end of play. Thus, in an anomalous situation, a number of events will still be observed after the decision has been taken by this awarding mechanism. However, the observed sequence of events will only be considered as anomalous when the confidence associated with its events by the weak classifier is high enough. If the reported events are correct, the only event that will create such anomaly is a bounce outside the play area. In the case of the ball is clearly out in singles tennis, the play will stop either immediately or after few ball events. In the doubles tennis, however, the tram lines are part of the play area and bounces in the tram lines will be seen as anomalous for an automatic system that is trained on singles tennis. (Note that the sequence of events that goes into the context classifier does not have multiple hypotheses except when there is uncertainty about bounce in or out (Eq. 1)). Through direct observation of singles tennis matches, we have established that the number of events reported subsequent to a clear bounce occurring outside of the legitimate play area does not appear to exceed four events. This is consequently our basis for classification of context.

4   Experiments

Experiments were carried out on data from two singles tennis matches and one doubles match. Training was done using 58 play shots of Women's final of the 2003 Australian Open tournament while 78 play shots of Men's final of the same tournament were used for validation. The test data is composed of 163 play shot of doubles Women's match of the 2008 Australian Open tournament. The data is manually annotated and 9 HMMs with 3 emitting states and 256 Gaussian mixture components per state modeling ball events were trained using the training data. The performance and parameterisation of these HMMs was optimised on the validation data. A window size of 7 observations is selected ($W = 3$ in Sec. 2.1). An accuracy of 88.73% event recognition was reached on the validation data, see figure 3(a). The last

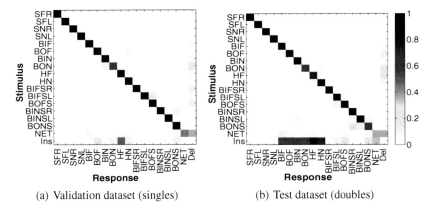

(a) Validation dataset (singles)　　　　　　　(b) Test dataset (doubles)

Fig. 3 Confusion matrices of recognised events, K. We use the same labels as [1].

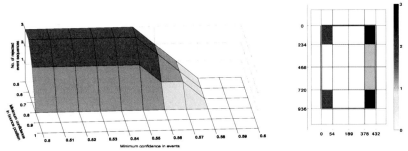

(a) Confidence of weak classifiers vs. no. rejected events (sin-　(b) Anomalies (doubles)
gles)

Fig. 4 (a) Number of event sequences of the validation set that contain errors with varied confidence thresholds and (b) Detected anomaly triggering events of the test set.

column of the matrix represents the number of deletions and the last row represents the number of insertions.

The confidence measures for the validation data were then used to find appropriate thresholds for rejecting sequences of events that are anomalous due to processing errors rather than genuine bounces out of the play area (Fig. 4(a)). The x-axis shows the minimum confidence reported on events up to the point of decision made about the event sequence while the y-axis shows the minimum confidence reported on bounces in or out the play area up to that point. The number of event sequences where the score decision is taken before the play ends are shown on the z-axis. It can be seen that a threshold of 0.8 in the bounce position confidence and 0.58 in the event recognition confidence lead to no false positives on sequences from singles. Applied on the doubles data, an accuracy of 83.18% event recognition was obtained using the parameters optimised on the singles data (Fig. 3(b)). The anomaly detector was able to detect 6 event sequences that contain anomaly, i.e. evident bounce in the tram lines followed by 5 or more events, out of a total of 21 anomaly sequences.

5 Data Association/Rule Updating

Having identified a discrete set of anomalous events in the manner indicated above, we proceed to an analysis of their import. A histogram of the detected anomalies superimposed on the court delineation obtained by extrapolation of detected horizontal and vertical court-line sets is given in figure 4(b). We also assume two axes of symmetry around the horizontal and vertical mid lines.

The determination of changes to play area definitions via events histogram is generally complex, requiring stochastic evaluation across the full lattice of possibilities [4]. However, in the absence of false positives, and with given the relatively complete sampling of the relevant court area it becomes possible to simplify the process. In particular, if we assume that only the main play area is susceptible to boundary redefinition, then the convex hull of the anomaly-triggering events is sufficient to uniquely quantify this redefinition.

In our case, this redefines the older singles play area coordinates $\{0,936,54,378\}$ to a new play area with coordinates $\{0,936,0,432\}$ shown in Fig. 4(b), where rectangles are defined by $\{$first row, last row, first column, last column$\}$, in inches. This identification of new expanded play area means that rules associated with activity in those areas now relate to the new area. Since the old area is incorporated within the novel area according to lattice inclusion, all of the detected anomalies disappear with the play area redefinition. In fact, the identified area $\{0,936,0,432\}$ corresponds exactly to the tennis doubles play area (i.e. the area incorporating the 'tram-lines').

6 Conclusions

We set out to implement an anomaly detection method in the context of sport video annotation, and to build upon it using the notion of *learning transfer* in order to incorporate the detected anomalies within the existing domain model in a conservative fashion.

In our experiments, the domain model consists in a fixed game rule structure applied to detected low-level events occurring within delineated areas (which vary for different game types). On the assumption that anomalies are due to the presence of a novel game domain, the problem is consequently one of determining the most appropriate redefinition of play areas required to eliminate the anomaly.

We thus applied an anomaly detection methodology to the problem of detecting the rule-incongruence involved in the transition from singles to doubles tennis videos, and proceeded to demonstrate how it may be extended so as to transfer learning from the one rule-governed structure to another via the redefinition of the main play area in terms of the convex hull of the detected anomalies. We thereby delineate two distinct rule domains or *contexts* within which the low-level action and event detectors and classifiers function. This was uniquely rendered possible by the absence of false positives in the anomaly detection; more stochastic methods would be required were this not the case.

References

1. Almajai, I., Kittler, J., de Campos, T., Christmas, W., Yan, F., Windridge, D., Khan, A.: Ball event recognition using HMM for automatic tennis annotation. In: Proc. Int. Conf. on Image Processing, Hong Kong, September 26-29 (2010) (in Press)
2. Burget, L., Schwarz, P., Matejka, P., Hannemann, M., Rastrow, A., White, C., Khudanpur, S., Hermansky, H., Cernock, J.: Combination of strongly and weakly constrained recognizers for reliable detection of OOVs. In: Proc. International Conference on Acoustics, Speech, and Signal Processing (ICASSP), pp. 4081–4084 (2008)
3. de Campos, T., Barnard, M., Mikolajczyk, K., Kittler, J., Yan, F., Christmas, W., Windridge, D.: An evaluation of bags-of-words and spatio-temporal shapes for action recognition. In: Proc. of the 10th IEEE Workshop on Applications of Computer Vision, Kona, Hawaii, January 5-6 (2011)
4. Khan, A., Windridge, D., de Campos, T., Kittler, J., Christmas, W.: Lattice-based anomaly rectification for sport video annotation. In: Proc. ICPR (2010)
5. Kittler, J., Christmas, W.J., Yan, F., Kolonias, I., Windridge, D.: A memory architecture and contextual reasoning for cognitive vision. In: Proc. SCIA, pp. 343–358 (2005)
6. Kläser, A., Marszałek, M., Schmid, C.: A spatio-temporal descriptor based on 3D-gradients. In: 19th British Machine Vision Conference, pp. 995–1004 (2008)
7. Kolonias, I.: Cognitive Vision Systems for Video Understanding and Retrieval. PhD thesis, University of Surrey (2007)
8. Tommasi, T., Caputo, B.: The more you know, the less you learn: from knowledge transfer to one-shot learning of object categories. In: British Machine Vision Conference (2009)
9. Wang, H., Ullah, M.M., Käser, A., Laptev, I., Schmid, C.: Evaluation of local spatio-temporal features for action recognition. In: 20th British Machine Vision Conference (2009)
10. Weinshall, D., Hermansky, H., Zweig, A., Luo, J., Jimison, H., Ohl, F., Pavel, M.: Beyond novelty detection: Incongruent events, when general and specific classifiers disagree. In: Advances in Neural Information Processing Systems (NIPS) (December 2009)
11. Yan, F., Christmas, W., Kittler, J.: Layered data association using graph-theoretic formulation with application to tennis ball tracking in monocular sequences. Transactions on Pattern Analysis and Machine Intelligence (2008)
12. Young, S., Kershaw, D., Odell, J., Ollason, D., Valtchev, V., Woodland, P.: The HTK Book Version 3.0. Cambridge University Press, Cambridge (2000)
13. Zimmermann, K., Svoboda, T., Matas, J.: Adaptive parameter optimization for real-time tracking. In: Proc. ICCV, Workshop on Non-rigid Registration and Tracking through Learning (2007)
14. Zimmermann, K., Svoboda, T., Matas, J.: Simultaneous learning of motion and appearance. In: The 1st International Workshop on Machine Learning for Vision-based Motion Analysis, Marseilles (2008), In Conjunction with ECCV
15. Zweig, A., Weinshall, D.: Exploiting object hierarchy: Combining models from different category levels. In: IEEE 11th International Conference on Computer Vision (2007)

Learning from Incongruence

Tomáš Pajdla, Michal Havlena, and Jan Heller

Abstract. We present an approach to constructing a model of the universe for explaining observations and making decisions based on learning new concepts. We use a weak statistical model, e.g. a discriminative classifier, to distinguish errors in measurements from improper modeling. We use boolean logic to combine outcomes of direct detectors of relevant events, e.g. presence of sound and presence of human shape in the field of view, into more complex models explaining the states in which the universe may appear. The process of constructing a new concept is initiated when a significant disagreement – incongruence – has been observed between incoming data and the current model of the universe. Then, a new concept, i.e. a new direct detector, is trained on incongruent data and combined with existing models to remove the incongruence. We demonstrate the concept in an experiment with human audio-visual detection.

1 Introduction

Intelligent systems compare their model of the universe, the "theory of the universe", with observations and measurements they make. The comparison of conclusions made by reasoning about well established building blocks of the theory with direct measurements associated with the conclusions allow to falsify [1] current theory and to invoke a rectification of the theory by learning from observations or restructuring the derivation scheme of the theory. It is the disagreement – incongruence – between the theory, i.e. derived conclusions, and direct observations that allows for developing a richer and better model of the universe used by the system.

Works [2, 3] proposed an approach to modeling incongruences between classifiers (called detectors in this work) which decide about the occurrence of concepts

Tomáš Pajdla · Michal Havlena · Jan Heller
Center for Machine Perception, Department of Cybernetics, FEE, CTU in Prague,
Technická 2, 166 27 Prague 6, Czech Republic
e-mail: {pajdla,havlem1,hellej1}@cmp.felk.cvut.cz

D. Weinshall, J. Anemüller, and L. van Gool (Eds.): DIRAC, SCI 384, pp. 119–127.
springerlink.com   © Springer-Verlag Berlin Heidelberg 2012

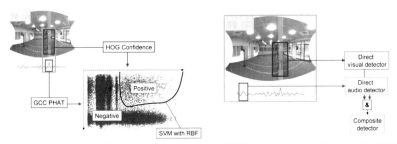

(a) The direct audio-visual (A∨)   (b) The composite audio-visual (A&V)
        human speaker detector   human speaker detector

Fig. 1 (a) The direct audio-visual human speaker detector constructed by training an SVM classifier with a RBF kernel in the two-dimensional feature space of GCC-PHAT values (x-axis) and pedestrian detection scores (y-axis) for different positive (*red circles*) and negative (*blue crosses*) manually labeled examples [6]. (b) The composite audio-visual human speaker detector accepts if and only if the direct visual detector AND the direct audio detector both accept (but possibly at different places) in the field of view. See [6] or an accompanied paper for more details.

(events) via two different routes of reasoning. The first way uses a single *direct* detector trained on complete, usually complex and compound, data to decide about the presence of an event. The alternative way decides about the event by using a *composite* detector, which combines outputs of several (in [2, 3] direct but in general possibly also other composite) detectors in a probabilistic (logical) way, Figure 1.

Works [2, 3] assume direct detectors to be independent, and therefore combine probabilities by multiplication for the "part-membership hierarchy", resp. by addition for the "class-membership hierarchy". Assuming trivial probability space with values 0 and 1, this coincides with logical AND and logical OR. Such reasoning hence corresponds to the Boolean algebra [4]. In the next we will look at this simplified case. A more general case can be analyzed in a similar way.

The theory of incongruence [2, 3] can be used to improve low-level processing by detecting incorrect functionality and repairing it through re-defining the composite detector. In this work we look at an example of incongruence caused by the omission of an important concept in an example of audio-visual speaker detection and show how it can be improved. Figure 1 and Table 1 illustrate a prototypical system consisting of alternative detectors, which can lead to a disagreement between the alternative outcomes related to an event.

Three direct detectors and one composite detector are shown in Figure 2(a). The direct detector of "Sound in view", the direct detector of "Person in view", the direct detector of "Speaker", and the composite detector of "Speaker" are presented. The composite detector was constructed as a logical combination of direct detectors evaluated on the whole field of view, hence not capturing the spatial co-location of sound and look events defining a speaker in the scene. See [6] or an accompanied paper for more details.

Table 1 Interpretation of agreement/disagreement of alternative detectors of the "Speaker" concept and their possible outcomes. See text, [2, 3].

	Direct	Composite	Possible reason
1	reject	reject	new concept or noisy measurement
2	reject	accept	incongruence
3	accept	reject	wrong model
4	accept	accept	known concept

Table 1 shows the four possible combinations of outcomes of the direct and composite "Speaker" detectors as analyzed in [3]:

The first row of the table, where none of the detectors accept, corresponds to no event, noise or a completely new concept, which has not been yet learned by the system. The last row of the table, when both detectors accept, corresponds to detecting a known concept.

The second row, when the "Speaker" composite detector accepts but the direct one remains negative, corresponds to the *incongruence*. This case can be interpreted as having a partial model of a concept, e.g. not capturing some important aspect like the spatial co-location if the composition is done by AND. Alternatively, it also can happen when the model of the concept is wrong such that it mistakenly allows some situations which are not truly related to the concept if the composition is done by OR.

The third row of the table, when the direct "Speaker" detector accepts but the composite one remains negative, corresponds to the *wrong model* case. Indeed, this case applies when the composite detector mistakenly requires some property which is not truly related to the concept if the composition is done by AND. It also happens when the composite detector has only a partial model of the concept, e.g. when it misses one of possible cases in which the concept should be detected, if the composition is done by OR.

We can see that the interpretation depends on how the composite detectors are constructed. Restricting ourselves to Horn clauses [7], i.e. to making the compositions by AND, we interpret the second and third rows as in [3]. Horn clauses are a popular choice since they allow efficient manipulation, and are used in PRO-LOG [8].

Assume to have a composite detector C constructed in the form of a Horn clause of direct detectors $D_1, D_2, \ldots, D_n$

$$D_1 \wedge D_2 \wedge \ldots D_n \rightarrow C \tag{1}$$

which means that C is active if and only if all D_i are active. For instance "Person in view" $\wedge$ "Sound in view" $\rightarrow$ "Speaker" is a Horn clause.

With this restriction, a detected incongruence can be understood as if the composite detector missed a term on the left hand side of the conjunction in the derivation rule, which is responsible for falsely rejected cases. It is easy to remedy this

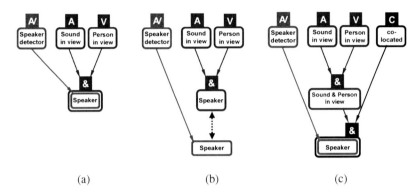

Fig. 2 (a) Two different detectors are executed in parallel. The direct A/ detector captures concisely all observed data involved in the training. The composite A&V detector aims at deriving ("explaining") the presence of a human speaker in the scene from primitive detections of human look a human sound. (b) Since the composite A&V detector does not properly capture co-location of human look and human sound to represent a true human speaker, a discrepancy – incongruence – appears when the A/ detector accepts but the A&V detector rejects. (c) An improved composite A&V&C detector uses a direct detector of space-time co-occurrence to better model (and hence detect correctly) human speakers.

situation by learning a new concept corresponding to the missing term in the conjunction from the incongruent examples.

There are many possibilities how to do it. A particularly simple way would be to add a single new concept "Co-located" to the conjunction, i.e.

$$\text{"Person in view"} \wedge \text{"Sound in view"} \wedge \text{"Co-located"} \to \text{"Speaker"} \qquad (2)$$

which would "push" the composite detector "down" to coincide with the direct detector, Figure 2(b).

Somewhat more redundant but still feasible alternative would be to add two more elements to the system as shown in Figure 2(c). A new composite detector "Sound & Person in view" could be established and combined with another newly introduced concept "Co-located" to update the model in order to correspond to the evidence. Although somewhat less efficient, this second approach may be preferable since it keeps concepts for which detectors have been established already.

As suggested above, the incongruence, i.e. the disagreement between the direct and the composite detectors, may signal that the composite detector is not well defined. We would like to use the incongruent data to learn a new concept, which could be used to re-define the composite detector and to remove the incongruence.

In the case of the speaker detector, the composite audio-visual detector has to be re-defined. A new "Sound & Person in view" concept has to be initiated. The composite audio-visual detector is disassociated from the "Speaker" concept, new "Sound & Person in view" concept is created and associated to the composite detector. This new concept will be greater than the "Speaker" concept. Next, a new

Fig. 3 The field of view is split into 20 segments (bins) and direct audio "Sound in view" detector (**A**) and direct video "Person in view" detector (**V**) are evaluated in each bin.

composite audio-visual "Speaker" detector is created as a conjunction of the composite "Sound & Person in view" detector and a new detector of an "XYZ" concept which needs to be trained using the incongruent data. The new composite detector is associated to the "Speaker" concept. The name of the "XYZ" concept can be established later based on its interpretation, e.g. as "Co-located" here.

2 Learning "Co-located" Detector

We will deal with the simplest case when the incongruence is caused by a single reason which can be modeled as a new concept. Our goal is to establish a suitable feature space and to train a direct detector deciding the "XYZ" concept using the congruent and incongruent data as positive and negative training examples respectively. As the values of audio and visual features have been used in direct audio and direct visual detectors already and our new concept should be as general as possible, we will use only boolean values encoding the presence of a given event in the 20 angular bins, Figure 3, [6].

First, two feature vectors of length 20 are created for each frame, one encoding the presence of audio events and the other one encoding the presence of visual events, and concatenated together in order to form a boolean feature vector of length 40.

Secondly, in order to find dependencies between the different positions of the events, the feature vector is lifted to dimension 820 by computing all possible products between the 40 values. The original feature vector:

$$x_1 \, x_2 \, x_3 \, \ldots \, x_{39} \, x_{40}$$

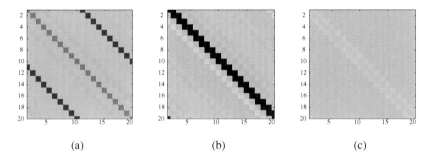

Fig. 4 Resulting weights for different pairs of values in the feature vector as obtained by SVM. Numbers on axes correspond to bin indices. Positive weights are denoted by red, zero weights by green and negative weights by blue color. (a) Audio × audio. (b) Audio × visual. (c) Visual × visual.

is transformed into:

$$x_1^2\, x_1 x_2\, x_1 x_3\, \ldots\, x_2^2\, x_2 x_3\, \ldots\, x_{39} x_{40}\, x_{40}^2.$$

When the original values are boolean, the quadratic monomials $x_1^2, x_2^2, \ldots$ have values equal to $x_1, x_2, \ldots$ and the monomials $x_1 x_2, x_1 x_3, \ldots$ have values equal to the conjunctions $x_1 \wedge x_2, x_1 \wedge x_3, \ldots$. SVM training over these vectors should reveal significant pairs of positions by assigning high weights to the corresponding positions in the lifted vector.

We use two sequences to construct the positive and negative training example sets. Each of these sequences is nearly 5 minutes long with a person walking along the line there and back with a loudspeaker placed near one of the ends of the line. During the first approx. 90 seconds, the walking person is speaking and the loudspeaker is silent rendering a congruent situation. During the next approx. 90 seconds, the walking person is silent and the loudspeaker is speaking causing an incongruent situation. In the last approx. 90 seconds, both the walking person and the loudspeaker are speaking which is congruent by our definition as we are able to find a bin with a speaker.

For each frame, the concatenated boolean feature vector is created. The boolean decisions for the 61 directional feature vectors from the audio detector [6] are transformed into the 20 values of the audio part of the feature vector using disjunction when more directional labels fall into the same bin. The visual part of the feature vector is initialized with 20 zeros and each confident human detection output by the visual detector changes the value belonging to the angular position of the center of the corresponding rectangle to one. Feature vectors belonging to frames which yielded a positive response from both the composite and the direct audio-visual detectors, i.e. congruent situation, are put into the positive set, those belonging to frames that were classified positively by the composite but negatively by the direct audio-visual detector, i.e. incongruent situation, are put into the negative set, and

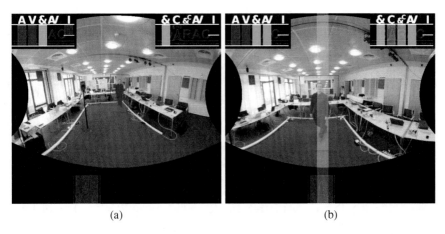

(a)   (b)

Fig. 5 (a) The original composite detector falsely detects a speaker in the scene with a silent person and sound generated by a loudspeaker at a different place (response & in the top left corner). The improved composite detector (response $\&^C$ in the top right corner) gives the correct result. (b) Both detectors are correct when a speaking person is present in the field of view.

those belonging to frames with a negative response from the composite audio-visual detector are discarded as such data cannot be used for our training.

As the loudspeaker position is fixed in our training sequences, we decided to remove the bias introduced by this fact by "rotating" the data around the bins, so each training example is used to generate 19 other training examples before lifting, e.g. a training example:

$$x_1 \, x_2 \, x_3 \, \ldots \, x_{20} \, x_{21} \, \ldots \, x_{39} \, x_{40}$$

is used to generate 19 additional examples:

$$x_2 \, x_3 \, x_4 \, \ldots \, x_1 \, x_{22} \, \ldots \, x_{40} \, x_{21}$$
$$x_3 \, x_4 \, x_5 \, \ldots \, x_2 \, x_{23} \, \ldots \, x_{21} \, x_{22}$$
$$\vdots$$
$$x_{20} \, x_1 \, x_2 \, \ldots \, x_{19} \, x_{40} \, \ldots \, x_{38} \, x_{39}.$$

Finally, the feature vectors of 60,320 positive and 41,820 negative examples were lifted and used to train a linear SVM detector [5].

The results shown in Figure 4 can be commented as follows. The most significant result is the dark red main diagonal in the $A \times V$ diagram, Figure 4(b), telling us that the positive examples have the audio and visual events in the same bin (or shifted by one bin as one of the neighboring diagonals is red too). Red square at $(1, 20)$ is a by-product of the "rotation" as neighboring bins can be separated to different ends of the view-field.

As can be seen in the $V \times V$ diagram, Figure 4(c), pairs of visual events are insignificant. The orange main diagonal in the $A \times A$ diagram, Figure 4(a), says that positive examples tend to contain more audio events. This is due to the fact that the only the situation with two audio events present in the training data was congruent, we had no training data with two loudspeakers speaking. The light blue adjacent diagonal is also an artifact of the direct audio-visual detector and "rotation".

To conclude, the just trained detector decides the position consistency of the audio and visual events, so a suitable name for the "XYZ" concept would be the "Co-located" concept. The resulting "Co-located" detector can be used to augment the initial composite "Speaker" detector, Figure 2(a), to produce new "Speaker" detectors, Figure 2(b,c). Figure 5(a) shows the original composite detector which falsely detects a speaker in the scene with silent person and sound generated by a loudspeaker at a different place (response & in the top left corner). The improved composite detector (response $\&^C$ in the top right corner) gives the correct results. In Figure 5(b), both detectors are correct when a speaking person is present in the field of view.

3 Conclusions

We have seen that incongruence can be used to indicate where to improve the model of the universe. It provided the training labels to data to pick up the right training set for constructing a detector explaining the missing concept in the model. Of course, we have just demonstrated the approach on the simplest possible realistic example with a single clause explaining the state of the world by a very few concepts.

The next challenge is to find an interesting, realistic but tractable problem leading to a generalization of the presented approach. It would be interesting to use more general logical formulas as well as to deal with errors in direct detectors.

Acknowledgements. This work was supported by the EC project FP6-IST-027787 DIRAC and by Czech Government under the research program MSM6840770038. We would like to acknowledge Vojtěch Franc for providing the large-scale SVM training code.

References

1. Popper, K.R.: The Logic of Scientific Discovery. Routledge, New York (1995)
2. Pavel, M., Jimison, H., Weinshall, D., Zweig, A., Ohl, F., Hermansky, H.: Detection and identification of rare incongruent events in cognitive and engineering systems. Dirac white paper, OHSU (2008)
3. Weinshall, D., et al.: Beyond novelty detection: Incongruent events, when general and specific classifiers disagree. In: NIPS 2008, pp. 1745–1752 (2008)
4. Halmos, P.R.: Lectures on Boolean Algebras. Springer, Heidelberg (1974)
5. Franc, V., Sonneburg, S.: Optimized cutting plane algorithm for large-scale risk minimization. Journal of Machine Learning Research 10, 2157–2232 (2009)

6. Pajdla, T., Havlena, M., Heller, J., Kayser, H., Bach, J.H., Anemüller, J.: Incongruence detection for detecting, removing, and repairing incorrect functionality in low-level processing. Research Report CTU–CMP–2009–19, Center for Machine Perception, K13133 FEE Czech Technical University (2009)
7. Wikipedia: Horn Clause (2010),
   http://en.wikipedia.org/wiki/Horn_clause
8. Wikipedia: PROLOG (2010), http://en.wikipedia.org/wiki/Prolog

Towards a Quantitative Measure of Rareness

Tatiana Tommasi and Barbara Caputo

Abstract. Within the context of detection of incongruent events, an often over-looked aspect is how a system should react to the detection. The set of all the possible actions is certainly conditioned by the task at hand, and by the embodiment of the artificial cognitive system under consideration. Still, we argue that a desirable action that does not depend from these factors is to update the internal model and learn the new detected event. This paper proposes a recent transfer learning algorithm as the way to address this issue. A notable feature of the proposed model is its capability to learn from small samples, even a single one. This is very desirable in this context, as we cannot expect to have too many samples to learn from, given the very nature of incongruent events. We also show that one of the internal parameters of the algorithm makes it possible to quantitatively measure incongruence of detected events. Experiments on two different datasets support our claim.

1 Introduction

The capability to recognize, and react to, rare events is one of the key features of biological cognitive systems. In spite of its importance, the topic is little researched. Recently, a new theoretical framework has emerged [7], that defines rareness as an incongruence compared to the prior knowledge of the system. The model has shown to work on several applications, from audio-visual persons identification [7] to detection of incongruent human actions [5].

A still almost completely unexplored aspect of the framework is how to react to the detection of an incongruent event. Of course, this is largely influenced by the task at hand, and by the type of embodiment of the artificial system under consideration: the type of reactions that a camera might have are bound to be different from the type

Tatiana Tommasi · Barbara Caputo
Idiap Research Institute, Centre Du Parc, Rue Marconi 19, P.O. Box 592,
CH-1920 Martigny, Switzerland
e-mail: {ttommasi,bcaputo}@idiap.ch

D. Weinshall, J. Anemüller, and L. van Gool (Eds.): DIRAC, SCI 384, pp. 129–136.
springerlink.com   © Springer-Verlag Berlin Heidelberg 2012

of actions a wheeled robot might take. Still, there is one action that is desirable for every system, regardless of their given task and embodiment: to learn the detected incongruent event, so to be able to recognize it correctly if encountered again in the future.

In this paper we propose a recently presented transfer learning algorithm [6] as a suitable candidate for learning a newly detected incongruent event. Our method is able to learn a new class from few, even one single labeled example by exploiting optimally the prior knowledge of the system. This would correspond, in the framework proposed by Weinshall et al, to transfer from the general class that has accepted. Another remarkable feature of our algorithm is that the internal parameter, that controls the amount of transferred knowledge, shows different behaviors depending on how similar the new class is to the already known classes. This suggests that it is possible to derive from this parameter a quantitative measure of incongruence for new detected events. Preliminary experiments on different databases support our claims.

2   Multi Model Transfer Learning

Given k visual categories, we want to learn a new $k + 1$ category having just one or few labeled data. We can use only the available samples and train on them, or we can take advantage of what already learned. The Multi model Knowledge Transfer algorithm (Multi-KT) addresses this latter scenario in a binary, discriminative framework based on LS-SVM [6]. In the following we describe briefly the Multi-KT algorithm. The interested reader can find more details in [6].

Suppose to have a binary problem and a set of l samples $\{\mathbf{x}_i, y_i\}_{i=1}^{l}$, where $\mathbf{x}_i \in \mathscr{X} \subset \mathbb{R}^d$ is an input vector describing the ith sample and $y_i \in \mathscr{Y} = \{-1, 1\}$ is its label. We want to learn a linear function $f(\mathbf{x}) = \mathbf{w} \cdot \phi(\mathbf{x}) + b$ which assigns the correct label to an unseen test sample $\mathbf{x}$. $\phi(\mathbf{x})$ is used to map the input samples to a high dimensional feature space, induced by a kernel function $K(\mathbf{x}, \mathbf{x}') = \phi(\mathbf{x}) \cdot \phi(\mathbf{x}')$ [2].

If we call $\mathbf{w}'_j$ the parameter describing the old models of already known classes $(j = 1, \ldots, k)$, we can write the LS-SVM optimisation problem slightly changing the regularization term [6]. The idea is to constrain a new model to be close to a weighted combination of pre-trained models:

$$\min_{\mathbf{w}, b} \frac{1}{2} \left\| \mathbf{w} - \sum_{j=1}^{k} \beta_j \mathbf{w}'_j \right\|^2 + \frac{C}{2} \sum_{i=1}^{l} \zeta_i (y_i - \mathbf{w} \cdot \phi(\mathbf{x}_i) - b)^2 . \tag{1}$$

Here $\boldsymbol{\beta}$ is a vector containing as many elements as the number of prior models k, and has to be chosen in the unitary ball, i.e. $\|\boldsymbol{\beta}\|_2 \leq 1$. Respect to the original LS-SVM, we are also adding the weighting factors ζ_i, they help to balance the contribution of the sets of positive (l^+) and and negative (l^-) examples to the data misfit term:

$$\zeta_i = \begin{cases} \frac{l}{2l^+} & \text{if } y_i = +1 \\ \frac{l}{2l^-} & \text{if } y_i = -1 . \end{cases} \tag{2}$$

With this new formulation the optimal solution is

$$\mathbf{w} = \sum_{j=1}^{k} \beta_j \mathbf{w}'_j + \sum_{i=1}^{l} \alpha_i \phi(\mathbf{x}_i) . \tag{3}$$

Hence $\mathbf{w}$ is expressed as a sum of the pre-trained models scaled by the parameters β_j, plus the new model built on the incoming training data.

An advantage of the LS-SVM formulation is that it gives the possibility to write the LOO error in closed form [1]. The LOO error is an unbiased estimator of the classifier generalization error and can be used for model selection [1]. A closed form for the LOO error can be easily written even for the modified LS-SVM formulation:

$$r_i^{(-i)} = y_i - \tilde{y}_i = \frac{\alpha_i}{\mathbf{G}_{ii}^{-1}} - \sum_{j=1}^{k} \beta_j \frac{\alpha'_{i(j)}}{\mathbf{G}_{ii}^{-1}}, \tag{4}$$

where $\alpha'_{i(j)} = \mathbf{G}_{(-i)}^{-1}[\hat{y}_1^j, \ldots, \hat{y}_{i-1}^j, \hat{y}_{i+1}^j, \ldots, \hat{y}_l^j, 0]^T$, $\hat{y}_i^j = (\mathbf{w}'_j \cdot \phi(\mathbf{x}_i))$ and $\tilde{y}_i$ are the LOO predictions. The $\mathbf{G}$ matrix is $[\mathbf{K} + \frac{1}{C}\mathbf{W}, \mathbf{1}; \mathbf{1}^T, 0]$, $\mathbf{K}$ is the kernel matrix, $\mathbf{W} = diag\{\zeta_1^{-1}, \zeta_2^{-1}, \ldots, \zeta_l^{-1}\}$, and $\mathbf{G}_{(-i)}$ is obtained when the ith sample is omitted in $\mathbf{G}$.

If we consider as loss function $loss(y_i, \tilde{y}_i) = \zeta_i \max[1 - y_i\tilde{y}_i, 0]$, to find the best $\boldsymbol{\beta}$ vector we need to minimise the objective function:

$$J = \sum_{i=1}^{l} \max\left[y_i\zeta_i\left(\frac{\alpha_i}{\mathbf{G}_{ii}^{-1}} - \sum_{j=1}^{k}\beta_j\frac{\alpha'_{i(j)}}{\mathbf{G}_{ii}^{-1}}\right), 0\right] \quad \text{s.t.} \quad \|\boldsymbol{\beta}\|_2 \le 1 . \tag{5}$$

3 Stability as a Quantitative Measure of Incongruence

An important property of Multi-KT is its stability. Stability here means that the behaviour of the algorithm does not change much if a point is removed or added. This notion is closely related to the LOO error, which is exactly calculated measuring the performance of the model every time a point is removed. From a practical point of view, this should correspond to a graceful decreasing of the variations in $\boldsymbol{\beta}$ as new samples arrive. This decrease of variations as the training data for the new class arrives should also be related to how difficult it is to learn it. Indeed, if the algorithm does not transfer much, we expect that $\boldsymbol{\beta}$ will stabilize slowly. This corresponds to the situation where the new class is very different from all the classes already learned– in other words, we expect that the stability of $\boldsymbol{\beta}$ is correlated to the rareness of the incoming class.

4 Experiments

This Section presents three set of experiments designed to test our claim that the stability of $\boldsymbol{\beta}$ is related to the rareness of the incoming class. We first show that,

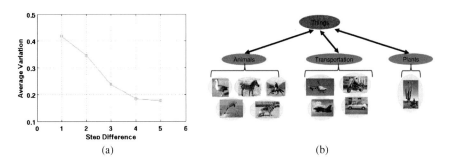

(a)   (b)

Fig. 1 (a) Norm of the difference between two β vectors correspondent two subsequent step in time. The norms are averaged both on the classes and on the splits; (b) Classes extracted from the Caltech-256 database: goose, zebra, horse, dophin, dog, helicopter, motorbike, fighter-jet, car-side, cactus.

as expected, β gets stable smoothly when the number of training samples grows (Section 4.1). We then explore how this behavior changes when considering prior knowledge related or unrelated to the new class. This is done first on an easy task (Section 4.2) and then in a more challenging scenario (Section 4.3).

For the experiments reported in Section 4.1 and 4.3 we used subsets of the Caltech-256 database [4] together with the features described in [3], available on the authors' website[1]. For the experiments reported in Section 4.2 we used the audio-visual database and features described in [7] using only the face images. All the experiments are defined as "object vs background" where the background corresponds respectively to the Caltech-256 clutter class and to a synthetically defined non-face, obtained scrumbling the face feature vector elements.

4.1 A Stability Check

As a first step we want to show that the variation in the β vector is small when the algorithm is stable. We consider the most general case of prior knowledge consisting of of a mix of related and unrelated categories. We therefore selected ten classes from the Caltech-256 database (see Figure: 1(b)). We run experiments ten times considering in turn one of the classes as the new one and all the other as prior knowledge. We defined 6 steps in time corresponding to a new sample entering the training set. For each couple of subsequent steps we calculated the difference between the obtained β vectors. Figure 1(a) shows the average norm of these differences and demonstrates that the algorithm stability does translate in a smooth decrease in the β vector of Multi-KT.

---

[1] http://www.vision.ee.ethz.ch/ pgehler/projects/iccv09/

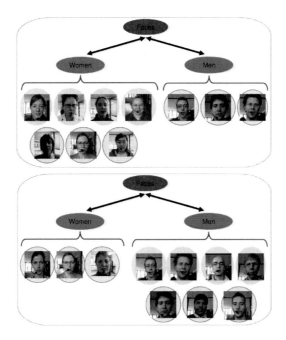

Fig. 2 Top: four women faces used as prior knowledge while three men and three women faces are considered in learning; Bottom: four men faces used as prior knowledge while three men and three women faces are considered in learning.

4.2 Experiments on Visual Data: Easy Learning Task

In the second set of experiments we dealt with the problem of learning male/female faces when prior knowledge consisted of only female/male faces. A scheme of the two experiments is shown in Figure 2.

For the first experiment, prior knowledge consisted of four women; the task was to learn three new men and three new women. Results are reported in Figure 3(a). The learning curves clearly indicate that the task becomes very easy when using the transfer learning mechanism: we obtain 100 % accuracy even with just one training sample, regardless of the gender. It is interesting to note that the information coming from the female face models is helpful for learning models of male faces. This is understandable, as they all are faces. Nevertheless, the difficulty in relying on faces of the opposite gender is still readable in Figure 3(b) which reports the norm of the differences between two β vectors for two subsequent steps in time.

We repeated the experiment using four men as prior knowledge for the task to learn the faces of three new men and three new women. Figure 4(a) show again that there is no significative difference between the two transfer learning curves obtained when learning man and woman faces, and they correspond both to 100 % accuracy. Looking at Figure 4(b) we notice that the β vector results more stable when learning a face of the same gender of those contained in the prior knowledge.

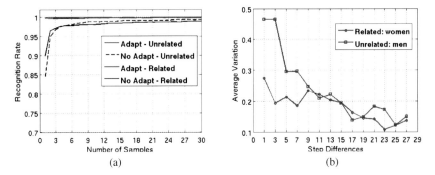

Fig. 3 Women as prior knowledge. (a) Classification performance as a function of the number of training images. The results shown correspond to average recognition rate considering each class out experiments repeated ten times. (b) Norm of the difference between two β vectors correspondent two subsequent step in time. The norms are averaged both on the classes and on the splits.

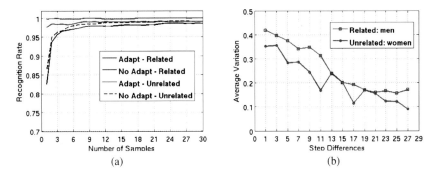

Fig. 4 Men as prior knowledge. (a) Classification performance as a function of the number of training images. The results shown correspond to average recognition rate considering each class out experiments repeated ten times. (b) Norm of the difference between two β vectors correspondent two subsequent step in time. The norms are averaged both on the classes and on the splits.

4.3   Experiments on Visual Data: Difficult Learning Task

In the third experiment we consider two different scenarios. In the first, we have a set of animals as prior knowledge and the task is to learn a new animal. In the second we have a mix of unrelated categories and the task is to learn a new one. From the point of view of transfer learning we expect the first problem to be easier than the second. Namely, in the first case only 1-2 labeled samples should be necessary, while in the second case the algorithm should need more samples.

To verify this hypothesis we extracted six classes from the Caltech-256 general category "Animal, land" and another group of six was defined picking each class

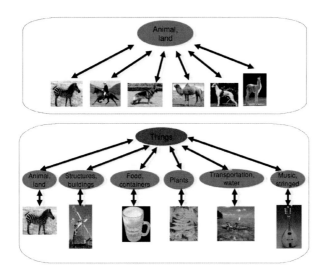

Fig. 5 Top: six classes from the Caltech-256 general category "Animal, land" (zebra, horse, dog, camel, llama, greyhound). Bottom: six classes extracted each form a general category of the Caltech-256 (zebra from "Animal, land", windmill from "Structures, building", beermug from "Food,containers", fern from "Plants", canoe from "Transportation, water" and mandolin from "Music, stringed").

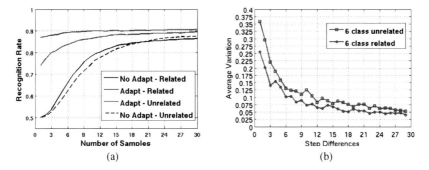

Fig. 6 (a) Classification performance as a function of the number of training images. The results shown correspond to average recognition rate considering each class out experiments repeated ten times. (b) Norm of the difference between two β vectors correspondent two subsequent step in time. The norms are averaged both on the classes and on the splits.

from a different general category (see Figure 5). Two different experiments were run: one with only the animal related classes, considering in turn 5 classes as known and one as new. The second, following the same setting on the six unrelated classes. Even if the two experiments were run separately, the non-transfer learning curve for the problems do not present a significant difference (see Figure 6(a)). This allow us to benchmark the corresponding results for learning with adaptation.

Figure 6(a) shows that when prior knowledge is not informative the algorithm needs more labeled data to learn the new class, demonstrating our initial intuition. In Figure 6(b) the corresponding norm of the differences between two $\boldsymbol{\beta}$ vectors for two subsequent steps in time is reported. We can compare the curves supposing to choose a treshold in the $\boldsymbol{\beta}$ variation: to reach $\Delta\boldsymbol{\beta} < 0.15$ it is necessary to have at least 3 samples when using related prior knowledge and 6 samples for unrelated prior knowledge. For $\Delta\boldsymbol{\beta} < 0.1$, 6 samples are required using related prior knowledge and 12 for unrelated, while to have $\Delta\boldsymbol{\beta} < 0.075$, 10 samples are needed using related prior knowledge and 18 for unrelated.

5   Conclusions

In this paper we addressed the problem of what action an artificial cognitive system can take, upon detection of an incongruent event. We argued that learning the new event from few labeled samples is one of the most general and desirable possible actions, as it does not depend on the embodiment of the system, nor its task. We showed how a recently introduced transfer learning algorithm could be used for this purpose, and also how its internal parameter regulating transfer learning could be used for evaluating the degree of incongruence of the new event. Future work will explore further this intuition, with the goal to derive a principled foundation for these results.

Acknowledgements.   This work was supported by the DIRAC (FP6-0027787) project.

References

1. Cawley, G.C.: Leave-one-out cross-validation based model selection criteria for weighted LS-SVMs. In: IJCNN (2006)
2. Cristianini, N., Shawe-Taylor, J.: An Introduction to Support Vector Machines. Cambridge University Press, Cambridge (2000)
3. Gehler, P., Nowozin, S.: Let the kernel figure it out: Principled learning of pre-processing for kernel classifiers. In: Proc. CVPR (2009)
4. Griffin, G., Holub, A., Perona, P.: Caltech 256 object category dataset. Technical Report UCB/CSD-04-1366, California Institue of Technology (2007)
5. Nater, F., Grabner, H., van Gool, L.: Exploiting simple hierarchies for unsupervised human behavior analysis. In: Proc. CVPR (2010)
6. Tommasi, T., Orabona, F., Caputo, B.: Safety in numbers: Learning categories from few examples with multi model knowledge transfer. In: Proc. CVPR (2010)
7. Weinshall, D., Hermansky, H., Zweig, A., Luo, J., Brgge Jimison, H., Ohl, F., Pavel, M.: Beyond novelty detection: Incongruent events, when general and specific classifiers disagree. In: Proc. NIPS (2008)

Part V
How Biological Systems Deal with Novel and Incongruent Events

Predictions and Incongruency in Object Recognition: A Cognitive Neuroscience Perspective

Helena Yardley, Leonid Perlovsky, and Moshe Bar

Abstract. The view presented here is that visual object recognition is the product of interaction between perception and cognition, a process budding from memories of past events. Our brain is proactive, in that it is continuously aiming to predict how the future will unfold, and this inherent ability allows us to function optimally in our environment. Our memory serves as a database from which we can form analogies from our past experiences, and apply them to the current moment. We are able to recognize an object through top-down and bottom-up processing pathways, which integrate to facilitate successful and timely object recognition. Specifically, it is argued that when we encounter an object our brain asks "what is this like" and therefore draws on years of experience, stored in memory, to generate predictions that directly facilitate perception. These feed-forward and feedback systems tune our perceptive and cognitive faculties based on a number of factors: predictive astuteness, context, personal relevance of the given event, and the degree to which its potential rarity differs from our original expectations. We discuss both computational and theoretical models of object recognition, and review evidence to support the theory that we do not merely process incoming information serially, and that during our attempts to interpret the world around us, perception relies on existing knowledge as much as it does on incoming information.

Helena Yardley
Martinos Center for Biomedical Imaging at Massachusetts General Hospital, USA
e-mail: helena.yardley@gmail.com

Leonid Perlovsky
Harvard University and Air Force Research Laboratory, Hanscom AFB, USA
e-mail: leonid@seas.harvard.edu

Moshe Bar
Martinos Center Biomedical Imaging at Massachusetts General Hospital,
Harvard Medical School, Charlestown, MA 02129, USA
e-mail: bar@mmr.mgh.harvard.edu

D. Weinshall, J. Anemüller, and L. van Gool (Eds.): DIRAC, SCI 384, pp. 139–153.
springerlink.com   © Springer-Verlag Berlin Heidelberg 2012

Introduction

Traditionally, perception and cognition are considered separately. By this main-stream view, perception pertains to the analysis of sensory input, while cognition is presumed to apply to higher-level processes that follow the interpretation of the input. But this boundary is superficial, as argued here, in that perception and cognition continuously interact with each other towards the goal of understanding our environment. There is an intimate relationship between perception and cognition, we perceive. This perceptual-cognitive interface provides the basis by which we

can experience, learn, and function more optimally. The result allows us to learn and form memories, and those memories subsequently allow us to create predictions about what might happen next.

Modeling this process mathematically by matching sensory data to memories has been attempted in artificial intelligence since the 1950s, but these attempts have had very limited success due to inadequate knowledge of the underlying neural mechanisms [71]. As an alternative, it is suggested that the human brain is proactively generating predictions about the relevant sights, sounds, and situations that we might encounter in the imminent future [3,4,44,12,63,10]. These predictions serve an important purpose, by preparing us and directing our attention to what is likely to happen, and alerting us to what is novel.

Predictions are an integral part of our everyday life, enabled by neural pathways of top-down and bottom-up processes of recognition that allow us to understand our world and the objects within it. For this review, we will focus

Fig. 1 Perception, cognition, memory, and the resulting predictions are feed-forward and feed-back systems, constantly changing the way we function in the future with the growing amount of situations we encounter.

on predictions in visual object and scene recognition. Our proposal is that when our brain encounters a new object, it does not attempt to recognize it by asking "what is this," which is inherently a bottom-up process, but rather by asking "what is this **like**" [3], which is a process that emphasizes experience and the role of memory in our understanding of the world around us. We continuously, consciously or not, search for analogies between new inputs and familiar representations stored in memory from experience. Finding such analogies not only facilitates recognition of what is in front of us, but also provides immediate access to a web of associations in memory that is the basis for the activation of predictions about likely upcoming situations. We will use this framework to later explain how the systems adjust to identify rare and incongruent aspects of our environment that happen to fall outside the predicted "net" of possible outcomes.

Predictions about the relevant future rely heavily upon our past memories; we are able to learn, and can therefore anticipate the future to some degree by drawing upon these memories and "lessons learned." In addition to memories from real experiences, our minds are also capable of mental simulations, or "imagining" the outcomes of imaginary events. For example, you might be swept up in a daydream during the lull of a late afternoon workday about what you would do should the building catch fire, and a swift evacuation was required. You would imagine how you would try to escape the building, help your fellow co-workers, or make a mental map of the nearest windows should the regular exits cease to be an option. These mental simulations then become part of your own memory database, and even though they are memories of events that have not yet happened, they could guide future actions as effectively as real memories.

At the center of this review is evidence to support the theory that we do not process our world simply by analyzing incoming information, but rather attempt to understand it by linking this incoming information to existing memories by way of analogy. When we encounter a novel stimulus (and all visual stimuli are novel to some extent, because we never experience the same situation twice under exactly the same conditions), our brains attempt to link this input with prior, similar experiences. In other words, we assess the global characteristics of, say, an object, and make a quick comparison as to how this object is likened to one that we are already familiar with, to help us identify the new object in question. By linking a novel input to a similar representation in memory (i.e. an analogy), we immediately gain access to a vast web of associations that are linked to this representation, and correspondingly to this novel input. Activating these associations, which are typically context-sensitive, is the mechanism that underlies prediction.

1 Object Recognition: Overview

The visual cortex is built in a hierarchical manner, which might be the primary reason why bottom-up processing has been the dominant dogma in thinking about visual processing for so many years. Visual perception begins with an image projected onto the retina, which is then relayed to the lateral geniculate nucleus (LGN). The LGN then projects to the primary visual cortex (V1), giving rise to the ventral pathway, which continues onward to the prestriate cortex (V2), to an area in the prelunate gyrus (V4), and ends in the temporal lobe where object recognition is presumably accomplished [37]. As information progresses through this visual processing hierarchy, neurons respond to increasingly complex stimuli. For instance, neurons located in the V1 area respond to very simple stimuli (e.g. orientation of lines) [49], while neurons in the temporal areas are selective for complex visual input, such as faces [72,89,85,58]. In addition, receptive field sizes (i.e. the size of the visual field to which a certain neuron is responsive) become progressively larger along the visual hierarchy, allowing processing of objects of larger magnitude and varying location [52].

Traditionally, visual object recognition has been taken as mediated by a hierarchical, bottom-up stream that processes an image by systematically analyzing its individual elements, and relaying the information to the next areas until the overall form and identity is determined. This idea has been challenged in recent years by theories and studies that are based on a more comprehensive model that combines bottom-up and top-down processing, which facilitates rapid and accurate object recognition as these "counter-streams" come to an agreement [54,90,42,15,31,56,1,6].

Top-down processing can be seen as making a quick and educated "guess," consulting with prior memories to activate simultaneous predictions about the identity of the input object, and finding an analogy of "What is this like?" For instance, seeing a dim and fuzzy picture of a hairdryer would allow you to quickly determine what the object may be (hairdryer, power drill, gun, etc), and what it definitely could not be (horse, car, toilet seat, etc.). In this way, we "prime" our systems for what to expect, and when our predictions are accurate, it facilitates faster recognition and further generation of association-based predictions. Once this global information is obtained, it is back-projected to the temporal cortex for collaboration in bottom-up processing [1]. A wealth of recent studies indicate that object identification is initiated and facilitated by top-down processes [48,17,60,32,6].

The extent of neural processing needed for object recognition is proportional to how congruent the input is with expectations, and the frequency with which the object has been encountered in the past. Previous neuroimaging studies have consistently shown a reduction in neural activity, particularly in the prefrontal cortex (PFC) and temporal cortex, for repeated stimuli [27,14,92,91,23]. This could be attributed to more efficient processing of a familiar stimulus, or the increased "synchrony" between top-down processing in the PFC and bottom-up processing in the temporal cortex [39,92,91,47,55,35,23,34,80,26,61]. A recent study by Ghuman et. al (2008) suggested that there is a relationship between cross-cortical communication and priming. It is possible that repetitions of a stimulus strengthen the connections between the nodes of this PFC-inferiotemporal (IT) network. Their results indicate that this feedback process is initiated in the PFC, which is then influencing processes in the temporal cortex. Top-down initiation of object recognition allows bottom-up process to function more effectively, and achieve object identification faster than a solely bottom-up process. In some instances it is even impossible to recognize an input object based on bottom-up processes exclusively.

One of the main criticisms of an exclusively bottom-up processing framework is that it presupposes that object recognition occurs only after the area at the tail end of the processing hierarchy has received and analyzed all the required input from earlier areas [1]. This scheme attributes substantially less importance to the parallel and feedback connections known to exist [78,16,65,74]. Object recognition is not unimodal; there are a number of other cues that lend themselves to a better understanding of one particular object, such as contextual cues, and our expectations about what we should encounter. In this review, we will examine the influence of top-down predictions in the process of object recognition, and explain how it might be integrated with other perceptual and cognitive processes to achieve visual object identification.

2 Mechanisms of Top-Down Activation

Top-down processing in object recognition can be seen as the brain's initial guess of "what is this like?" and is greatly dependent on the global physical properties of the stimulus. Instead of processing individual details, this process relies on the "gist" of the object. Results from previous studies have indicated that people perceive the low spatial frequency (LSF) components of an object prior to perceiving high spatial frequency (HSF) components [25,50,66], potentially providing the means for triggering predictions in a top-down fashion [81,71,6,73,67]. According to one specific framework (Bar, 2003), this LSF information is projected rapidly from the early visual areas directly to the prefrontal cortex possibly via the dorsal magnocellular pathway. This blurred delineation sparks a cascade of events that begin to generate predictions about the identity of the object in the orbitofrontal cortex (OFC) and downstream to the visual cortex (Figure 2) [6]. The process of LSF-to-HSF recognition proposed in [1] and demonstrated in [6] shares several principles with Carpenter & Grossberg's (1987) ART framework, Perlovsky's dynamic logic model of cognition (2006) and others. Attempting to process the details of an object sequentially would take substantially longer without top-down guidance. Also, in cases where an immediate assessment is needed (i.e. if you saw someone in your periphery coming at you), being able to rely on this LSF information could mean a faster reaction time, and, in extreme cases, the difference between life and death. This process allows us to make a rapid, and possibly subconscious "guess" as to what an object in our field of view may be.

Neurophysiological studies have shed light on the issue, providing evidence for the proposition that top-down processes initiate object recognition [48,17,60,32,6]. Bar et. al. (2006) found that object recognition based on LSF information elicited differential activity in the left OFC 50 ms earlier than areas in the temporal cortex when trying to identify an object. From the initial activity in the OFC, the sequence of recognition progressed to the left fusiform gyrus (180 ms), then to the left fusiform gyrus (215 ms). The neural activity was most robust at 130 ms from the presentation of the object, and remained statistically significant for approximately 40 ms. In addition, trials carried out with an easily recognizable object elicited substantially less robust early OFC activity than the masked trials. This further supports the proposition that top-down processing initiates predictions for facilitating object recognition.

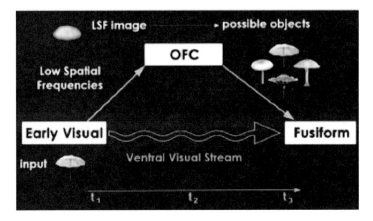

Fig. 2 Information from the LSF components of an image is projected from early visual cortex to the OFC, possibly via the magnocellular pathway. As the OFC generates predictions about the object's identity, information is simultaneously processed in a bottom-up manner along the relatively slower ventral visual pathway. Information from the two visual pathways are integrated to accomplish object recognition. (Adapted from Bar et. al, 2006.)

Other studies have shown evidence to support the presence of cortical and subcortical shortcuts in general, including projections of coarse information from the thalamus to the amygdale [57], from V1 and V2 to V4 and the posterior part of IT [65], and between V4 and TE [33]. Additionally, studies have shown that the PFC receives magnocellular projections [8,79,40]. Specifically, the magnocellular pathway has direct connections from as early as visual area V2 to the dorsolateral PFC, and from the ventral area V4, directly to the ventrolateral PFC [78,8]. Therefore, these data provide evidence of the existence of the neural infrastructure required for the rapid projection of objects containing LSF from early visual areas to the PFC [1]. This information is then projected back to the ventral visual cortex where top-down and bottom-up processes may be integrated. This feedback system allows more efficient processing by considerably narrowing the number of possible object identities that need to be considered. In one mathematical model of this interaction between top-down and bottom-up signals [71], it was argued that vague LSF top-down contents are essential for overcoming combinatorial computational complexity that had plagued artificial intelligence and other mathematical attempts at object recognition in the past. Top-down processing provides a framework from which bottom-up processing can engage more efficiently. But in addition to the predictions that can be generated about an object's identity based on its global "gist" properties, a significant source of predictability can be gleaned from the context in which objects appear.

3 Objects in Context: Overview

A scene is contextually coherent if it is composed of items that frequently appear together within that context. Seeing a street scene prompts predictions about what other objects might also be encountered: a mailbox, cars, buildings, people. These predictions are triggered by stored associations that are based on the frequency with which we encounter these objects together on a regular basis, and these context-driven predictions allow us to save valuable processing power by inferring what objects are present, rather than having to attend specifically to each object within the scene. These predictions are activated rapidly, and can be triggered by LSF global information, just like we proposed with single objects. It has indeed been shown thoroughly that LSF images of typical contexts are highly informative about the context [88,68] and thus can serve the purpose of triggering predictions reliably. This predictive association might be initiated using Bayesian inference methods, based on priming, attention, and expectation of scales (e.g. size, distances) [88,53], and are further reinforced through cellular-level mechanisms such as long-term potentiation (LTP) [29] and Hebbian-based learning [46].

There are instances of conditionalized associative activation that hint to the fact that associations are not activated automatically. An image of a book is associated with a number of other objects, and our brain does not know if we are in a library, and need to activate corresponding associations such as rows of book shelves, chairs, desks, computers, and librarians, or if we are at home and need to activate, instead, the representation of a coffee table, a couch, and a television. It has been proposed [2,34] that our brain initially activates all these associations simultaneously, but once more information about the context (e.g. library) becomes available, it suppresses associations from the irrelevant context (e.g. home), and keeps online only the associations of the appropriate context frame.

Aside from objects that are related to each other contextually because they typically appear together, objects can be related to each other in various other ways. For example, a toaster and a microwave are contextually related, but two different brands of microwave are different types of the same basic-level concept and are thus semantically related; still, the simple visual components of a microwave and a mailbox, although not contextually related, are perceptually related. These various relationships are stored differently in different areas of the brain. Recent evidence has supported the theory that cortical representations of objects might be grouped by physical properties in early visual cortex [41,59,84], by basic-level categories in the anterior temporal cortex [76,45,28], by contextual relations in the parahippocampal cortex (PHC) [5] and by semantic relations in the PFC [36]. We maintain multiple different representations for the same object, each of which is geared towards and deployed for dedicated purposes.

Computational attempts to identify what constitute a coherent contextual scene have been undertaken since the 1960's and, similar to object recognition, have proven to be challenging. The mathematical and conceptual challenges associated with these topics have the same limiting factor: the multitude of conditional combinations a scene can contain. There are many objects in the field of view in any given direction, in any given moment. Which of them are relevant to which context? The human mind solves this problem so effortlessly that it is difficult to appreciate its computational complexity, which is even more involved than the previously discussed complexity of object recognition. Again, the gist-like nature of top-down projections have been fundamental in reducing computational

complexity, helping to understand the brain mechanism of top-down signal interaction that facilitates the recognition of objects in context [2,70].

In addition to objects that commonly co-occur in the same scene, or that are
conceptually related to each other in some way, Biederman et al. (1982) ex-
pounded on five types of object relations that characterize the rules that govern a
scene's structure and their influence on perception: familiar size (known relative
size of objects), position (typical spatial orientation), probability (frequency with
which you have encountered this object within this scene), support (objects tend to
be physically supported, not floating), and interposition (objects in the foreground
will occlude what is behind them) [13]. These rules help to form our *a priori*
knowledge about how the world should be, and summons our attention when objects appear to be incongruent with our expectations. In this way, scenes that follow these rules facilitate faster and more efficient processing, particularly when
one's predictions are congruent with the contextual frame and the object's identity. To a large extent, then, the process of object identification is dependent upon
how the object is related to its surroundings. [69,22,51,21,2].

The growing amount of scenes we encounter, our accumulated experience, allows us to "prime" for object recognition. This sort of priming can be seen as
"predictive coding" [30], and it involves a matching process between our top-
down predictions and bottom-up feature detection along the cortical visual systems [30,64,77,35,82]. One theory of why priming facilitates faster object recognition is that these "predictions" help to direct attention [75], but beyond attention,
which tells us where to expect something, the predictions also often contain "gist"
information of what to expect in that location. Therefore, a prediction is typically
a combination of space and physical features that together facilitate our perception
and action.

4 Rare Events and "Surprises"

A key concept when thinking about how we detect rare and incongruent events in
our environment is to realize that by generating a prediction about what might be,
the brain also implicitly predicts what might *not* be. A prediction activates certain
expectations, and by doing so inhibits, actively or not, alternatives that are unlikely to be relevant. As discussed, we typically have an underlying expectation of
what we will encounter, and these expectations help work synergistically with
feedback and feed-forward systems to identify objects within our environment. It
is a process that relies on memory, predictions, and collaborations among a number of different structures. But what happens when our predictions are incongruent
with the visual input? We live in a complex world, and in order to flourish in our
environment, we must possess the ability to adjust to a variety of situations flexibly, and react to them appropriately. In addition to being able to respond flexibly
when our expectations have been violated to some degree, prediction error provides one of the prime sources for learning [87,24]. The systems we have mentioned have a role in how this discrepancy is realized and learned from. We are
particularly interested here in the detection of rare events within the context of
prediction error, and how our neural systems adjust to process the dissonance encountered within our environment.

There are two mechanisms that modulate how rare events are detected and processed: attention and expectation. Attention prioritizes how stimuli are processed based on personal relevance, and expectations constrain visual interpretation based on prior likelihood [83]. We typically don't attend strongly to mundane objects within our environment, unless they are of some importance to us at that moment, though we may still perceive them on some lower level. Only when there is something out of the ordinary do we take closer notice. Our brains allow us to ignore, to some degree, the objects that are predictable, and activate visual object recognition regions more strongly when an object falls outside our pool of predicted objects. This creates a gradient of neural responses corresponding to how different the input material is from the original expectation, as well as how emotionally salient the information is (see Fig. 3). The frequency with which we encounter the same situation aids in developing an internal statistical probability of the most likely outcome, and these regularities continuously construct causal models of the world [43]. When the present events violate our predictions, the process of object recognition adjusts to reconcile the differences between what was expected and what is presented.

Neural Activity in OFC: *Incongruence vs. Saliency*

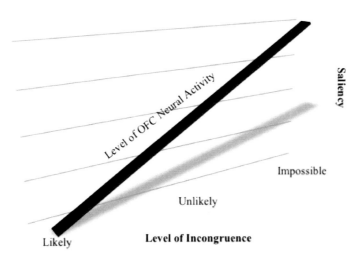

Fig. 3 The degree of neural activity associated with the process of object recognition is dependent upon how much the rare event differs from our expectations, as well as how much the mismatch between the event and our expectations is personally relevant. This effect is most pronounced in the OFC, where top-down predictions are thought to originate in the process of object recognition.

The information that is congruent with our expectations is processed differently than that of a rare event. Implicit in the definition of learning, learning occurs when we are confronted with *novel* information, and have to make sense of it. The identification of some novelty means that a rare event has occurred. We hold a

general expectation of how our environment should be, and rare events are identified when top-down predictions are mismatched with the incoming bottom-up information in a meaningful manner. This inconsistency is then noted, and for-
warded to memory to avoid prediction error in the future [35,30], which is the
basis of how we learn. At the moment that we are presented with a rare event, the
stimulus' discordant properties within the scene alert us to the conflict. This rare
event is realized when an object's individual characteristics conflict with our own
expectations of the world. As information travels up the visual processing hierarchy, each lower-level processing center interprets incoming data, a process that is
presumably guided by and compared with the top-down predictions. When an "error" in prediction has been perceived, the higher-level systems generate a new
prediction in accordance with the lower level information, and process the object's
features further until these two cognitive-perceptual streams come to an agreement
[77]. There are a number of other factors that further influence this circuitry, including the degree of deviation from our expectation, the personal relevance of the
deviation, as well as whether the situation poses a threat to our safety.

Emotions play a large part in perception as well. Every perceivable object and
event we encounter elicit either a positive, negative, or a neutral reaction, whether
it be a fleeting thought or a small physiological reaction. This discernment is made
by comparing the present moment to how a similar situation has affected us in the
past. The OFC integrates our external perceptions along with our internal bodily
ongoings to create a contextually specific representation of the situation, and its
personal relevance [7,9,95,20,62,38,11]. Current research has shown that emotionally salient stimuli initiate activity in the limbic and paralimbic systems [86].
There is strong activity in the amygdala, which has been traditionally seen as the
"fear center," when presented with a rare event that has the potential for personal
harm (e.g. a ninja perched upon the counter in your periphery). Sometimes the
presence of the rare event itself provides enough uncertainty, and therefore potential harm, to engage the amygdala.

5   Conclusions

Visual object recognition is typically a result of interplay between perception and
cognition. Science has only begun to uncover the mechanisms by which we perceive and comprehend our world. Our brain is proactive in generating predictions
about the relevant future, to better prepare us for what is to come next. We *learn*
when we encounter rare and novel events, and store this novelty for future reference to be better prepared for a similar event in the future. Rare events themselves
are approached and processed differently based upon our current state of mind and
needs. In one moment, we are actively concerned with a particular outcome, and
therefore attend strongly, and react strongly to the resulting situation. In the next
moment, the same outcome is no longer personally relevant, and therefore that
particular rarity goes unnoticed. Our sensory brain is an ever-changing entity, constantly acquiring new knowledge, computing statistical probabilities about our environment, and adapting to a changing environment. Our main purpose is survival,
and we are therefore hard-wired accurately and efficiently detect and reconcile

events that violate our expectations. Our neural circuitry enables us to thrive in our respective environments, no matter what rare events we may encounter.

Acknowledgments. Work on this chapter was supported by NIH, NSF, DARPA and the Air Force.

References

1. Bar, M.: A cortical mechanism for triggering top-down facilitation in visual object recognition. Journal of Cognitive Neuroscience 15, 600–609 (2003)
2. Bar, M.: Visual objects in context. Nature Reviews Neuroscience 5(8), 617–629 (2004)
3. Bar, M.: The proactive brain: Using analogies and associations to generate predictions. Trends in Cognitive Sciences 11(7), 280–289 (2007)
4. Bar, M. (ed.): Predictions: A universal principle in the operation of the human brain (Introduction). Theme issue: Predictions in the brain: using our past to generate a future. Philosophical Transactions of the Royal Society B: Biological Sciences (2009)
5. Bar, M., Aminoff, E.: Cortical analysis of visual context. Neuron 38(2), 347–358 (2003)
6. Bar, M., Kassam, K.S., et al.: Top-down facilitation of visual recognition. Proc. Natl. Acad. Sci. USA 103(2), 449–454 (2006)
7. Barbas, H.: Organization of cortical afferent input to orbitofrontal areas in the rhesus monkey. Neuroscience 56(4), 841–864 (1993)
8. Barbas, H.: Anatomic basis of cognitive-emotional interactions in the primate prefrontal cortex. Neuroscience and Biobehavioral Reviews 19(3), 499–510 (1995)
9. Barbas, H.: Connections underlying the synthesis of cognition, memory, and emotion in primate prefrontal cortices. Brain Research Bulletin 52(5), 319–330 (2000)
10. Barbey, A.K., Krueger, F., et al.: Structured event complexes in the medial prefrontal cortex support counterfactual representations for future planning. Philos. Trans. R Soc. Lond. B Biol. Sci. 364(1521), 1291–1300 (2009)
11. Barrett, L.F., Bar, M.: See it with feeling: affective predictions during object perception. Philos. Trans. R Soc. Lond. B Biol. Sci. 364(1521), 1325–1334 (2009)
12. Barsalou, L.W.: Simulation, situated conceptualization, and prediction. Philosophical Transactions of the Royal Society B: Biological Sciences Special Theme Issue (2009)
13. Biederman, I., Mezzanotte, R.J., et al.: Scene perception: detecting and judging objects undergoing relational violations. Cognit. Psychol. 14(2), 143–177 (1982)
14. Buckner, R., Goodman, J., et al.: Functional-anatomic correlates of object priming in humans revealed by rapid presentation event-related fMRI. Neuron 20, 285–296 (1998)
15. Bullier, J.: Integrated model of visual processing. Brain Research Reviews 36, 96–107 (2001)
16. Bullier, J., Nowak, L.G.: Parallel versus serial processing: New vistas on the distributed organization of the visual system. Current Opinion in Neurobiology 5, 497–503 (1995)
17. Cardin, V., Friston, K.J., et al.: Top-down modulations in the visual form pathway revealed with dynamic causal modeling. Cereb. Cortex 21(3), 550–562 (2011)
18. Carmichael, S.T., Price, J.L.: Limbic connections of the orbital and medial prefrontal cortex in macaque monkeys. Journal of Comparative Neurology 363(4), 615–641 (1995)

19. Carpenter, G.A., Grossberg, S.: Neural dynamics of category learning and recognition: Attention, memory consolidation, and amnesia. In: Davis, J., Newburgh, R., Wegman, E. (eds.) Brain Structure, Learning, and Memory, pp. 233–290. Erlbaum, Hillsdale (1987)

20. Cavada, C., Company, T., et al.: The anatomical connections of the macaque monkey orbitofrontal cortex. A review. Cerebral Cortex 10(3), 220–242 (2000)

21. Chun, M.M.: Contextual cueing of visual attention. Trends Cogn. Sci. 4(5), 170–178 (2000)

22. Cutler, B.L., Penrod, S.D.: Context reinstatement and eyewitness identification. In: Davies, G.M., Thomson, D.M. (eds.) Memory in context: Context in memory. John Wiley & Sons Ltd, Chichester (1988)

23. Dale, A., Liu, A.K., et al.: Dynamic statistical parametric mapping: Combining fMRI and MEG for high-resolution imaging of cortical activity. Neuron 26(1), 55–67 (2000)

24. den Ouden, H.E., Friston, K.J., et al.: A dual role for prediction error in associative learning. Cereb. Cortex 19(5), 1175–1185 (2009)

25. DeValois, R.L., DeValois, K.K.: Spatial vision. Oxford Science Publications, New York (1988)

26. Dhond, R.P., Buckner, R.L., Dale, A.M., Marinkovic, K., Halgren, E.: Spatiotemporal maps of activity underlying word generation and their modification during repetition priming. J. Neurosci. 21, 3564–3571 (2001)

27. Dobbins, I.G., Schnyer, D.M., et al.: Cortical activiy reductions during repetition priming can result from rapid response learning. Nature 428(6980), 316–319 (2004)

28. Downing, P.E., Jiang, Y., et al.: A cortical area selective for visual processing of the human body. Science 293(5539), 2470–2473 (2001)

29. Dudai, Y.: The neurobiology of memory. Oxford University Press, Oxford (1989)

30. Egner, T., Monti, J.M., et al.: Expectation and surprise determine neural population responses in the ventral visual stream. J. Neurosci. 30(49), 16601–16608 (2010)

31. Engel, A.K., Fries, P., et al.: Dynamic predictions: oscillations and synchrony in top-down processing. Nature Reviews Neuroscience 2(10), 704–716 (2001)

32. Esterman, M., Yantis, S.: Perceptual expectation evokes category-selective cortical activity. Cereb. Cortex 20(5), 1245–1253 (2010)

33. Felleman, D.J., Van Essen, V.C.: Distributed hierarchical processing in primate visual cortex. Cerebral Cortex 1, 1–47 (1991)

34. Fenske, M.J., Aminoff, E., et al.: Top-down facilitation of visual object recognition: Object-based and context-based contributions. Progress in Brain Research 155, 3–21 (2006)

35. Friston, K.: A theory of cortical responses. Philos. Trans. R. Soc. Lond. B. Biol. Sci. 360(1456), 815–836 (2005)

36. Gabrieli, J.D., Poldrack, R.A., et al.: The role of left prefrontal cortex in language and memory. Proceedings of the National Academy of Sciences, USA 95(3), 906–913 (1998)

37. Horton, J.C., Sinich, L.C.: A new foundation for the visual cortical hierarchy. In: Gazzaniga, M. (ed.) The Cognitive Neurosciences, 3rd edn., ch. 17, pp. 233–243 (2004)

38. Ghashghaei, H., Barbas, H.: Pathways for emotion: interactions of prefrontal and anterior temporal pathways in the amygdala of the rhesus monkey. Neuroscience 115, 1261–1279 (2002)

39. Ghuman, A., Bar, M., Dobbins, I.G., Schnyer, D.M.: The effects of priming on frontal-temporal communication. Proceedings of the National Academy of Science 105(24), 8405–8409 (2008)
40. Goldman-Rakic, P.S., Porrino, L.J.: The primate mediodorsal (MD) nucleus and its projection to the frontal lobe. Journal of Comparative Neurology 242(4), 535–560 (1985)
41. Grill-Spector, K., Kourtzi, Z., et al.: The lateral occipital complex and its role in object recognition. Vision Research 41(10-11), 1409–1422 (2001)
42. Grossberg, S.: How does a brain build a cognitive code? Psychological Review 87(1), 1–51 (1980)
43. den Ouden, H.E.M., Friston, K.J., Daw, N.D., McIntosh, A.R., Stephan, K.E.: A Dual Role for Prediction Error in Associative Learning. Cereb. Cortex 19(5), 1175–1185 (2009); first published online (September 26, 2008), doi:10.1093/cercor/bhn161
44. Hawkins, J., George, D., et al.: Sequence memory for prediction, inference and behaviour. Philos. Trans. R Soc. Lond. B Biol. Sci. 364(1521), 1203–1209 (2009)
45. Haxby, J.V., Gobbini, M.I., et al.: Distributed and overlapping representations of faces and objects in ventral temporal cortex. Science 293(5539), 2425–2430 (2001)
46. Hebb, D.O.: The organization of behavior. Wiley, New York (1949)
47. Henson, R.N.: Neuroimaging studies of priming. Progress in Neurobiology 70(1), 53–81 (2003)
48. Hirschfeld, G., Zwitserlood, P.: How vision is shaped by language comprehension–top-down feedback based on low-spatial frequencies. Brain Res. 1377, 78–83 (2011)
49. Hubel, D.H., Wiesel, T.N.: Ferrier lecture. Functional architecture of macaque monkey visual cortex. Proc. R Soc. Lond. B Biol. Sci. 198(1130), 1–59 (1977)
50. Hughes, H.C., Nozawa, G., et al.: Global precedence, spatial frequency channels, and the statistics of natural images. Journal of Cognitive Neuroscience 8(3), 197–230 (1996)
51. Intraub, H.: The representation of visual scenes. Trends Cogn. Sci. 1(6), 217–222 (1997)
52. Kandel, E.R., Schwartz, J.H., et al.: Principles of neural science. Elsevier, New York (1991)
53. Kersten, D., Mamassian, P., et al.: Object perception as Bayesian inference. Annu. Rev. Psychol. 55, 271–304 (2004)
54. Kosslyn, S.M.: Image and Brain. MIT Press, Cambridge (1994)
55. Kveraga, K., Ghuman, A.S., et al.: Top-down predictions in the cognitive brain. Brain and Cognition 65, 145–168 (2007)
56. Lamme, V.A.F., Roelfsema, P.R.: The distinct modes of vision offered by feedforward and recurrent processing. Trends in Neuroscience 23, 571–579 (2000)
57. LeDoux, J.E.: The emotional brain. Simon & Schuster, New York (1996)
58. Logothetis, N.K., Sheinberg, D.L.: Visual object recognition. Annu. Rev. Neurosci. 19, 577–621 (1996)
59. Malach, R., Levy, I., et al.: The topography of high-order human object areas. Trends Cogn. Sci. 6(4), 176–184 (2002)
60. Malcolm, G.L., Henderson, J.M.: Combining top-down processes to guide eye movements during real-world scene search. J. Vis. 10(2), 4, 1–11 (2010)
61. Marinkovic, K., Dhond, R.P., et al.: Spatiotemporal dynamics of modality-specific and supramodal word processing. Neuron 38(3), 487–497 (2003)

62. Mesulam, M.: Behavioral neuroanatomy: large-scale networks, association cortex, frontal syndromes, the limbic system, and hemispheric specializations. In: Mesulam, M. (ed.) Principles of Behavioral and Cognitive Neurology, 2nd edn., pp. 1–120. Oxford University Press, New York (2000)

63. Moulton, S.T., Kosslyn, S.M.: Imagining predictions: Mental imagery as mental simulation. Philosophical Transactions of the Royal Society B: Biological Sciences Special Theme Issue (2009)

64. Mumford, D.: On the computational architecture of the neocortex. II. The role of cortico-cortical loops. Biol. Cybern. 66(3), 241–251 (1992)

65. Nakamura, H., Gattass, R., et al.: The modular organization of projections for areas V1 and V2 to areas V4 and TEO in macaques. Journal of Neuroscience 13(9), 3681–3691 (1993)

66. Navon, D.: Forest before trees: The precedence of global features in visual perception. Cognitive Psychology 9, 1–32 (1977)

67. Neisser, U.: Cognitive psychology. Appleton-Century –Crofts, New York (1967)

68. Oliva, A., Torralba, A.: Modeling the shape of the scene: A holistic representation of the spatial envelope. International Journal of Computer Vision 42(3), 145–175 (2001)

69. Palmer, S.E.: The effects of contextual scenes on the identification of objects. Memory and Cognition 3, 519–526 (1975)

70. Perlovsky, L.I., Ilin, R.: Grounded Symbol. In: The Brain, Computational Foundations For Perceptual Symbol System. Webmed Central PSYCHOLOGY 2010, vol. 1(12), pp. WMC001357 (2010)

71. Perlovsky, L.I.: Toward Physics of the Mind: Concepts, Emotions, Consciousness, and Symbols. Phys. Life Rev. 3(1), 22–55 (2006)

72. Perrett, D.I., Hietanen, J.K., et al.: Organization and functions of cells responsive to faces in the temporal cortex. Philosophical Transactions of the Royal Society of London, B 335, 23–30 (1992)

73. Peyrin, C., Michel, C.M., et al.: The neural substrates and timing of top-down processes during coarse-to-fine categorization of visual scenes: a combined fMRI and ERP study. J. Cogn. Neurosci. 22(12), 2768–2780 (2010)

74. Porrino, L.J., Crane, A.M., et al.: Direct and indirect pathways from the amygdala to the frontal lobe in rhesus monkeys. Journal of Comparative Neurology 198(1), 121–136 (1981)

75. Posner, M.I., Snyder, C.R., et al.: Attention and the detection of signals. J. Exp. Psychol. 109(2), 160–174 (1980)

76. Puce, A., Allison, T., Asgari, M., Gore, J.C., McCarthy, G.: Differential sensitivity of human visual cortex to faces, letterstrings, and textures: a functional magnetic resonance imaging study. Journal of Neurosciences 16(16), 5205–5215 (1996)

77. Rao, R.P., Ballard, D.H.: Predictive coding in the visual cortex: a functional interpretation of some extra-classical receptive-field effects (see comments). Nature Neuroscience 2(1), 79–87 (1999)

78. Rempel-Clower, N.L., Barbas, H.: The laminar pattern of connections between prefrontal and anterior temporal cortices in the rhesus monkey is related to cortical structure and function. Cerebral Cortex 10(9), 851–865 (2000)

79. Russchen, F.T., Amaral, D.G., et al.: The afferent input to the magnocellular division of the mediodorsal thalamic nucleus in the monkey, Macaca fascicularis. Journal of Comparative Neurology 256(2), 175–210 (1987)

80. Schnyer, D.M., Dobbins, I.G., Nicholls, L., Schacter, D., Verfaellie, M.: Rapid Decision Learning Alters the Repetition N400 Components in Left Frontal and Temporal Regions: Evidence from MEG Recordings During Repetition Priming. Society for Neuroscience, Washington, D.C. (2005)
81. Schyns, P.G., Oliva, A.: From blobs to boundary edges: Evidence for time- and spatial-dependent scene recognition. Psychological Science 5(4), 195–200 (1994)
82. Spratling, M.W.: Predictive coding as a model of biased competition in visual attention. Vision Res. 48(12), 1391–1408 (2008)
83. Summerfield, C., Egner, T.: Expectation (and attention) in visual cognition. Trends Cogn. Sci. 13(9), 403–409 (2009)
84. Tanaka, K.: Neuronal mechanisms of object recognition. Science 262, 685–688 (1993)
85. Tanaka, K.: Representation of Visual Features of Objects in the Inferotemporal Cortex. Neural Netw. 9(8), 1459–1475 (1996)
86. Taylor, S.F., Phan, K.L., et al.: Subjective rating of emotionally salient stimuli modulates neural activity. Neuroimage 18(3), 650–659 (2003)
87. Tobler, P.N., O'Doherty, J.P., et al.: Human neural learning depends on reward prediction errors in the blocking paradigm. J. Neurophysiol. 95(1), 301–310 (2006)
88. Torralba, A.: Contextual priming for object detection. International Journal of Computer Vision 53(2), 153–167 (2003)
89. Tovee, M.J., Rolls, E.T., et al.: Translation invariance in the responses to faces of single neurons in the temporal visual cortical areas of the alert macaque. J. Neurophysiol. 72(3), 1049–1060 (1994)
90. Ullman, S.: Sequence seeking and counter streams: A computational model for bidirectional information flow in the visual cortex. Cerebral Cortex 1, 1–11 (1995)
91. Wig, G.S., Grafton, S.T., et al.: Reductions in neural activity underlie behavioral components of repetition priming. Nat. Neurosci. 8(9), 1228–1233 (2005)
92. Zago, L., Fenske, M.J., et al.: The rise and fall of priming: How visual exposure shapes cortical representations of objects. Cereb. Cortex 15, 1655–1665 (2005)

Modulations of Single-Trial Interactions between the Auditory and the Visual Cortex during Prolonged Exposure to Audiovisual Stimuli with Fixed Stimulus Onset Asynchrony

Antje Fillbrandt and Frank W. Ohl

Abstract. The perception of simultaneity between auditory and visual stimuli is of crucial importance for audiovisual integration. However, the speeds of signal transmission differ between the auditory and the visual modalities and these differences have been shown to depend on multiple factors. To maintain the information about the temporal congruity of auditory and visual stimuli, flexible compensation mechanisms are required. Several behavioral studies demonstrated that the perceptual system is able to adaptively recalibrate itself to audio-visual temporal asynchronies [33,94]. Here we explored the adaptation to audio-visual temporal asynchronies at the cortical level. Tone and light stimuli at the same constant stimulus-onset-asynchrony were presented repeatedly to awake, passively listening, Mongolian gerbils. During stimulation the local field potential was recorded from electrodes implanted into the auditory and the visual cortices. The dynamics of the interactions between auditory and visual cortex were examined using the Directed Transfer Function (DTF; [42]). With increasing number of stimulus repetitions the averaged evoked response of the Directed Transfer Function exhibited gradual changes in amplitude. A single-trial analysis indicated that the adaptations observed in the all-trial average were due to modulations of the amplitude of the single-trial DTFs but not to alterations in the trial-to-trial dynamics of DTF peaks.

Antje Fillbrandt
Leibniz-Institute for Neurobiology, Magdeburg, Germany
e-mail: antje.fillbrandt@ifn-magdeburg.de

Frank Ohl
Leibniz-Institute for Neurobiology, Magdeburg, Germany
Institute for Biology, University of Magdeburg, Magdeburg, Germany
e-mail: frank.ohl@ifn-magdeburg.de

D. Weinshall, J. Anemüller, and L. van Gool (Eds.): DIRAC, SCI 384, pp. 155–180.
springerlink.com   © Springer-Verlag Berlin Heidelberg 2012

1   Introduction

Several studies have demonstrated the crucial role of temporal stimulus congruity in the binding of multisensory information (e.g. [61,8]). Given the differences in physical and neuronal transmission times of auditory and visual signals, the question arises how synchronization of multisensory information is achieved in the brain. An increasing number of studies indicate that temporal perception remains plastic throughout life-time: when stimuli from different sensory modalities are presented repeatedly at a small constant temporal onset asynchrony after a while their temporal disparity is perceived as being diminished in the conscious experience. This chapter explores whether the synchronization dynamics between the primary auditory and visual cortex adapt flexibly to constant timing of auditory and visual stimuli. We applied a rodent preparation designed to mimic relevant aspects of classical experiments in humans on the recalibration of temporal-order judgment.

1.1   The Speed of Transmission of Signals Is Modality Specific

Apparently, precise information about the temporal congruity of multisensory information is not readily available in the nervous system. From the point in time a single event causes an auditory and a visual signal to the point in time a certain brain area is activated by these signals, the information about their relative timing is blurred by different speeds of transmission of the two signals in various ways. The first temporal disparities in signal propagation arise already outside the brain from the different velocities of sound and light. At the receptor level sound transduction in the ear is faster than phototransduction in the retina (see [28], for a detailed review). The minimum response latency for a bright flash, ca. 7 ms, is nearly the same in rods and cones [18,41,71]. But with low light intensities the rod-driven response might take as long as 300 ms [5,6]. In contrast, the transduction by the hair cells of the inner ear is effectively instantaneously via direct mechanic linkage (about 10 microseconds, [19,20,21,22]).

   Next, the duration of the transmission of auditory and visual signals depends on the length of the nerves used for their transmission [92,37]. The relationship of transmission delays between sensory modalities is further complicated by the fact that in each modality processing speed seems to be modulated by both the detailed physical stimulus characteristics, like stimulus intensity [97], visual eccentricity [65,49], etc., and additionally by subjective factors, like attention (e.g., [70].

1.2   Simultaneity Constancy

The ability to perceive stimuli as simultaneous despite their different transmission delays has been termed simultaneity constancy [49]. Several studies demonstrated that human beings were able to compensate for temporal lags caused by variances in spatial distance [26,80,49,2]. Interestingly, the compensation also worked when distance cues were presented only to a single modality. In the study of Sugita and Suzuki (2003) only visual distance cues were used; Alais and Carlile (2005) varied only cues for auditory distance perception. The question which cues are essential to induce a lag compensation are still a matter of an ongoing debate as there are also several studies failing to find evidence for a similar perceptual compensation [78,53,3,40].

1.3 Temporal Recalibration

The transmission delays of auditory and visual signals depend on multiple factors and cannot be described by simple rules. One way to deal with this complexity could be that the compensation mechanisms remain plastic throughout lifetime so that they can flexibly adapt to new sets of stimuli and their typical transmission delays.

Temporal recalibration to stimuli-onset-asynchronies of multimodal stimuli has been demonstrated in several studies [33,94,63,40,47]. In these studies, experimental paradigms typically start with an adaptation phase with auditory and visual stimuli being presented repeatedly over several minutes, consistently at a slight onset asynchrony of about zero to 250 milliseconds. In a subsequent behavioral testing phase auditory and visual stimuli are presented at various temporal delays and usually their perceived temporal distance is assessed by a simultaneity judgment task (subjects have to indicate whether the stimuli are simultaneous or not) or a temporal order judgment task (subjects have to indicate which of the stimuli they perceived first).

Using these procedures temporal recalibration could be demonstrated repeatedly: the average time one stimulus had to lead the other in order for the two to be judged as occurring simultaneously, the point of subjective simultaneity (PSS), was shifted in the direction of lag used in the adaptation phase [33,94]. For example, if sound was presented before the light in the adaptation phase, in the testing phase the sound stimulus has to be presented earlier in time than it had to before the adaptation, in order to be regarded as having occurred simultaneously with the light stimulus.

In addition, in several studies an increase in the Just Notable Difference was observed (JND, smallest temporal interval between the two stimuli needed for the participants in a temporal order task to be able to judge correctly which of the stimuli was presented first on 75% of the trials) [33,63].

1.4 Outlook on Experiments

We applied an experimental paradigm resembling closely the previously described human studies on temporal recalibration: here we presented auditory and visual stimuli repeatedly to awake, passively perceiving, Mongolian gerbils at a constant temporal onset asynchrony of 200 ms.

The neural mechanisms underlying temporal recalibration have not yet been investigated in detail. The idea that temporal recalibration works on an early level of processing is quite attractive: More accurate temporal information is available at early stages as the different processing delays of later stages have not yet been added. But there are also reasons to believe that recalibration works at later levels: recalibration effects are usually observed in conscious perception.

In the last decades the primary sensory cortices have repeatedly been demonstrated to be involved in multisensory interactions (e.g. [16,14,9,46,62]. In the current explorative study we start to search for neural mechanisms of recalibration at the level of primary sensory cortex. We implanted one electrode into the primary auditory cortex and one electrode into the visual cortex of Mongolian gerbils and during stimulation local field potentials were recorded in the awake animals.

There is accumulating evidence that the synchronization between brain areas might play an important role in crossmodal integration [11,12]. Directional influences of the auditory and visual cortex were analyzed by using the Directed Transfer Function (DTF) [42] on the local field potential data. Our main question of interest was whether the interaction patterns between auditory and visual cortex changed with increasing presentation number of asynchronous of auditory and visual stimuli.

A further important question in the investigation of multisensory integration is how unimodal areas can be integrated while maintaining the specialization within the areas. Nonlinear models on complex brain dynamics suggest that this aim might be achieved by constantly changing states of partial coordination between different brain areas [23,82,29]. To address this issue we additionally investigated the single-trial dynamics of cross-cortical interactions by characterizing the trial-to-trial variability of the DTF of short data windows.

2   Methods

2.1   Animals

Data were obtained from 8 adult male Mongolian gerbils (*Meriones unguiculatus*). All animal experiments were surveyed and approved by the animal care committee of the Land Sachsen-Anhalt.

2.2   Electrodes

Electrodes were made of stainless steal wire (diameter 185 μm) and were deinsulated only at the tip. The tip of the reference electrodes was bent into a small loop (diameter 0.6 mm). The impedance of the recording electrodes was 1.5 MΩ (at 1 kHz).

2.3   Animal Preparation and Recording

Electrodes were chronically implanted under deep ketamine anesthesia (xylazine, 2 mg/100 g body wt ip; ketamine, 20 mg/100 g body wt ip). One recording electrode was inserted into the right primary auditory cortex and one into the right visual cortex, at depths of 300 μm, using a microstepper. Two reference electrodes were positioned onto the dura mater over the region of the parietal and the frontal cortex, electrically connected and served as a common frontoparietal reference. After the operation, animals were allowed to recover for one week before the recording sessions began. During the measurements the animal was allowed to move freely in the recording box (20 × 30 cm). The measured local field potentials from auditory and visual cortex were digitized at a rate of 1000 Hz.

2.4 Stimuli

Auditory and visual stimuli were presented at a constant intermodal stimulus onset asynchrony of 200 ms. The duration of both the auditory and the visual stimuli was 50 ms and the intertrial interval varied randomly between one and two seconds with a rectangular distribution of intervals in that range. Acoustic stimuli were tones presented from a loudspeaker located 30 cm above the animal. The tone frequency was chosen for each individual animal to match the frequency that evoked in preparatory experiments the strongest amplitude of local field potential at the recording site within the tonotopic map of primary auditory cortex [67,68]. The range of the frequencies used reached from 250 Hz to 4 kHz with the peak level of the tone stimuli varying between 60 dB (low frequencies) and 48 dB ((high frequencies), measured by a Bruel und Kjaer Sound Level Meter Type). Visual stimuli were flashes presented from an LED-lamp (9.6 cd/m^2) located at the height of the eyes of the animal.

2.5 Experimental Protocol

To be able to examine both short-term and long-term adaptation effects animals were presented with asynchronous stimuli for ten sessions with 750 stimulus presentations at each session. For four animals the auditory stimuli were presented first, for the remaining four animals the visual stimuli were presented first.

2.6 Data Preprocessing

The local field potential (LFP) time series of each trial were analyzed from 1 s before to 1 s after the first stimulus. The LFP data of this time period were detrended, separately for each trial and each channel. In addition, the temporal mean and the temporal standard deviation of the time period were determined for each trial and for each channel and used for z-standardization. Amplifier clippings as they resulted from movement of the animals were identified by visual inspection. Only artifact-free trials were included into the analysis (about 70-90 % of the trials).

2.7 The Directed Transfer Function: Mathematical Definition

Directional influences between the auditory and the visual cortex were analysed in single trials by estimating the Directed Transfer Function (DTF) ([42,44] for comparison of the performance of the DTF with other spectral estimators see also

[52,4]). The Directed Transfer Function is based on the concept of Granger causality. According to this concept, one time series can be called causal to a second one if its values can be used for improving the prediction of the values of the second time series measured at later time points. This basic principle is typically mathematically represented in the formalism of autoregressive models (AR-models).

Let $X_1(t)$ be the time series data from a selectable channel one and $X_2(t)$ the data from a selectable channel 2:

$$X_1(t) = \sum_{j=1}^{p} A_{1 \to 1}(j) X_1(t-j) + \sum_{j=1}^{p} A_{2 \to 1}(j) X_2(t-j) + E \tag{1}$$

$$X_2(t) = \sum_{j=1}^{p} A_{1 \to 2}(j) X_1(t-j) + \sum_{j=1}^{p} A_{2 \to 2}(j) X_2(t-j) + E \tag{2}$$

Here, the $A(j)$ are the autoregressive coefficients at time lag j, p is the order of the autoregressive model and E the prediction error. According to the concept of Granger causality, in (1) the channel X_2 is said to have a causal influence on channel X_1 if the prediction error E can be reduced by including past measurement of channel X_2 (for the influence of the channel X_1 on the channel X_2 see (2)).

To investigate the spectral characteristics of the interchannel interaction the autoregressive coefficients in (1) were Fourier-transformed; the transfer matrix was then obtained by matrix inversion:

$$\begin{pmatrix} A_{1 \to 1}(f) A_{2 \to 1}(f) \\ A_{1 \to 2}(f) A_{2 \to 2}(f) \end{pmatrix}^{-1} = \begin{pmatrix} H_{1 \to 1}(f) H_{2 \to 1}(f) \\ H_{1 \to 2}(f) H_{2 \to 2}(f) \end{pmatrix} \tag{3}$$

where the component of the $A(f)$ matrix are:

$$A_{l \to m}(f) = 1 - \sum_{j=1}^{p} A_{l \to m}(j) e^{-i2\pi f j} \tag{4} \quad \text{when } l = m$$

with l being the number of the transmitting channel and m the number of the receiving channel

$$A_{l \to m}(f) = 0 - \sum_{j=1}^{p} A_{l \to m}(j) e^{-i2\pi f j} \tag{5} \quad \text{otherwise.}$$

The Directed Transfer Function (DTF) for the influence from a selectable channel 1 to a selectable channel 2, DTF $_{1 \to 2}$, is defined as:

$$nDTF_{1 \to 2}(f) = \left| H_{1 \to 2}(f)^2 \right| \tag{6}$$

In the case of only two channels, the DTF measures the predictability of the frequency response of a first channel from a second channel measured earlier in time. When, for example, X_1 describes the LFP from the auditory cortex, X_2 the LFP form the visual cortex and the amplitude of the $nDTF_{1 \to 2}$ has high values in the beta-band, it means that we are able to predict the beta-response of the visual cortex from the beta-response of the auditory cortex measured earlier in time. There are several possible situations of crosscortical interaction that might underlie modulation of DTF amplitudes (see for example [44,17,25]. See the Discussion section for more details.

2.8 Estimation of the Autoregressive Models

We fitted bivariate autoregressive models to LFP time series from auditory and visual cortex using the Burg method as this algorithm has been shown to provide accurate results [57,45,76]. We partitioned the time series data of single trials into 100-ms time windows which were stepped at intervals of 5 ms through each trial from one second before the first stimulus to one second after the first stimulus. Models were estimated separately for each time window of the single trials. Occasionally, the covariance matrix used for estimation of the AR-coefficients turned out to be singular or close to singular, in these rare cases the whole trial was not analysed any further.

In the present study we used a modal order of 8, the sampling rate of 1000 Hz was used for model estimation. The model order was determined by the Akaike Information Criterion [1]. After model estimation, the adequacy of the model was tested by analyzing the residuals [56]. Using this model order, the auto- and cross-covariance of the residuals was found to have values between 0.001 to 0.005 % of the auto and crosscovariance of the original data (data averaged from two animals here). In other words, the model was able to capture most of the covariance structure contained in the data. When DTFs were computed from the residuals, the single-trial spectra were almost flat, indicating that the noise contained in the residuals was close to white noise.

The estimation of AR-models requires the normality of the process. To analyze to which extent the normality assumption was fulfilled in our data, the residuals were inspected by plotting them as histograms and, in addition, a Lillie-test was computed separately for the residuals of the single data windows. In about 80% of the data windows the Lillie test confirmed the normality assumption.

A second requirement for the estimation of the autoregressive models is the stationary of the time series data. Generally, this assumption is better fulfilled with small data windows [24], though it impossible to tell in advance at which data window a complex system like the brain will move to another state [31].

A further reason why the usage of small data windows is recommendable is that changes in the local field potential are captured at a higher temporal resolution. The spectral resolution of low frequencies does not seem to be a problem for small data windows when the spectral estimates are based on AR-models (for a mathematical treatment of this issue, see for example [57], p. 199f).

Using a high sampling rate ensures that the number of data points contained in the small time windows is sufficient for model estimation. For example when we used a sampling rate of 500 Hz instead of 1000 Hz to estimate models from our time windows of 100 ms, the covariance of the residuals increased signaling that the estimation has become worse (the autocovariance of the residuals of the auditory and visual channels at 1000 Hz were about 10 % of the auto- and cross-covariance of the auditory and visual channels at 500 Hz). Importantly, when inspecting the spectra visually they seemed to be quite alike, indicating that AR-models are robust to an extent to a change in sampling rate. When using a data window of 200 ms with the same sampling rate of 500 Hz the model estimation improved (the covariance of the residuals was 20 % to 40 % of the covariance of a model with a window of 100 ms), but at the expense of the temporal resolution.

2.9   Normalization of the DTF

[42] suggested normalization of the Directed Transfer Function relative to the structure which sends the signal, i.e. for the case of the directed transfer from the auditory channel to the visual channel:

$$nDTF_{A \to V}(f) = \frac{\left| H_{A \to V}(f)^2 \right|}{\sum_{M=1}^{k} \left| H_{M \to V}(f) \right|^2} \tag{7}$$

In the two-channel case the $DTF_{A \to V}$ is divided by the sum of itself and the spectral autocovariance of the visual channel. Thus, when using this normalization the amplitude of the $nDTF_{A \to V}$ depends on the influence of the auditory channel on itself and, reciprocally, the amplitude of the $nDTF_{V \to A}$ is dependent on the influence of the visual channel on itself. This is problematic in two ways: First, we cannot tell whether differences between the amplitude of the $nDTF_{A \to V}$ and the amplitude of the $nDTF_{V \to A}$ are due to differences in normalization or to differences in the strengths of crosscortical influences. Second, analysis of our data has shown that the auditory and the visual stimulus influenced both the amplitude of the local field potential and the spectral autocovariance of both auditory and the visual channels. Thus it is not clear whether changes in the amplitude of the nDTF after stimulation signal changes in the cross cortical interaction or changes in spectral autocovariance of the single channels.

As the non-normalized DTF is difficult to handle due to large differences in the amplitudes at different frequencies we normalized the DTF in the following way:

$$nDTF_{A \to V}(f) = \frac{DTF_{A \to V}(f)}{\left(\sum_{1}^{n\_session} \sum_{1}^{n\_trials} \sum_{1}^{n\_windows} DTF_{A \to V}(f) \right) / (n\_windows * n\_trials * n\_session)} \tag{8}$$

with n_windows being the number of time windows of the prestimulus interval per trial,n_trials the number of trials per session, and n_session the number of sessions.

Hence, the amplitude of the DTF estimated for each single time window of the single trials was divided by the average of the DTF of all time windows taken from the 1-second- prestimulus interval of the single trials of all sessions.

2.10   Statistical Testing

We assessed the statistical significance of differences in the amplitude of the nDTF using the bootstrap technique (e.g. [27]) in order not be bound to assumptions about the empirical statistical error distribution of the nDTF (but see [25], for an investigation of the statistical properties of the DTF). The general procedure was as follows: First, bootstrap samples were drawn from the real data under the assumption that the null hypothesis was true. Then for each bootstrap sample a chosen test statistic was computed. The values of the test statistic from all bootstrap samples formed a distribution of values of the test statistic under the assumption of the null hypothesis. Next, we determined from the bootstrap distribution of the test statistic the probability of finding values equal or larger than the empirically observed one by chance. If this value was below the preselected significance level the null hypothesis was rejected.

More specifically, in our first bootstrap test we wanted to test the hypothesis whether the nDTF has higher amplitude values in the poststimulus interval than in the prestimulus interval. Under the assumption of the null hypothesis the nDTF-amplitude values of the pre- and the poststimulus interval should not be different from each other. Thus, pairs of bootstrap samples were generated by taking single trial nDTF-amplitude values at random but with replacement from the pre- and from the poststimulus interval. For each of the sample pairs the amplitudes were averaged across trials and the difference between the averages was computed separately for each pair. This procedure of drawing samples was repeated 1000 times getting a distribution of differences between the average amplitudes. The resulting bootstrap distribution was then used to determine the probability of the real amplitude difference of the averages between the pre- and the poststimulus interval under the assumption of the null hypothesis.

In a second bootstrap test we assessed the significance of the slope of a line fitted to the data by linear regression analysis. We used the null hypothesis that the predictor variable (here the number of stimulus presentations) and the response variable (here the nDTF amplitude) are independent from each other. We generated bootstrap samples by randomly pairing the values of the predictor and observer variables. For each of these samples a line was fitted by linear regression analysis and the slope was computed obtaining a distribution of slope values under the null hypothesis.

3   Results

3.1   Stimulus-Induced Changes in the Single–Trial nDTF, Averaged across All Trials of All Sessions

For a first inspection of the effect the audiovisual stimulation had on the nDTF from auditory to visual cortex ($nDTF_{A \to V}$) and from visual to auditory cortex ($nDTF_{V \to A}$) we averaged nDTF amplitudes across all single trials of all sessions, separately for each time window from one second before to one second after the first stimulus. Figure 1 shows time-frequency plots of the nDTF$_{A \to V}$ (figure 1A) which describes the predictability of the frequency response of the visual cortex based on the frequency response of the auditory cortex and the nDTF$_{V \to A}$ (figure 1B) which describes the predictability of the frequency response of the auditory cortex based on the frequency response of the visual cortex. Data is shown both for animals receiving the tone stimulus first (figure 1,left) for animals receiving the light stimulus first (figure 1, right) from 200 ms before the first stimulus to 1 s after the first stimulus here. Note that the abscissa indicates the start of a time window (window duration: 100 ms), so the data from time windows at 100 ms before the first stimulus are already influenced by effects occurring after the presentation of the first stimulus.

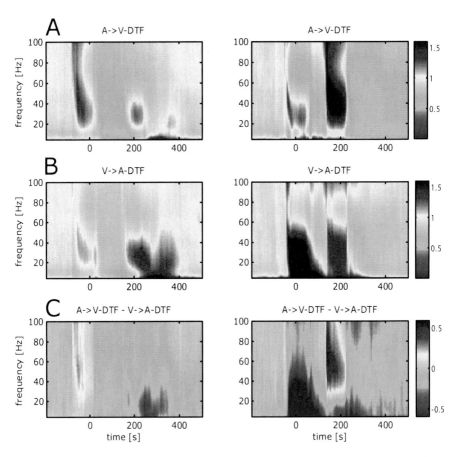

Fig. 1 A & B: Normalized nDTF$_{A \to V}$ (A) and nDTF$_{V \to A}$ (B) averaged across all trials from all sessions, separately for time windows from -0.2 to 0.9 s after the start of the first stimulus. Left: animal receiving first the tone. Right: animal receiving first the light. C: Difference between averages (nDTF$_{A \to V}$ - nDTF$_{V \to A}$).

The significance of the observed changes in the nDTF-amplitude was assessed separately for each animal using the student's t-test based on the bootstrap technique (see Methods). More precisely, we tested whether the amplitudes of the nDTF averaged across trials at different time points after the presentation of the first stimulus were significantly different from the nDTF amplitude of the prestimulus interval, averaged across trials and time from -1000 ms to 100 ms before the first stimulus. To compare the relative amplitudes of the $nDTF_{A \to V}$ and the $nDTF_{V \to A}$, we tested whether the difference of the amplitudes of AV- and $nDTF_{V \to A}$ averaged across trials at different time points after the presentation of the first stimulus were significantly different from the difference of the amplitudes of $nDTF_{A \to V}$ and $nDTF_{V \to A}$ of the prestimulus interval. In the following we will describe only peaks of the amplitudes of nDTF which deviated significantly ($p < 0.01$) from the average amplitude of prestimulus interval.

3.1.1   Animals Receiving First the Light, Then the Tone Stimulus (VA-Animals)

At first sight the response of the $nDTF_{A \to V}$ resembled closely the response of the $nDTF_{V \to A}$. In animals receiving first the light stimulus and then the tone stimulus we observed two prominent positive peaks in both the $nDTF_{A \to V}$ (figure 1A, right) and the $nDTF_{V \to A}$ (figure 1B, right), the first one after the light stimulus started at about - 20 ms and the second one after the tone stimulus began at about 151 ms. After the second peak the amplitude of the $nDTF_{A \to V}$ and the $nDTF_{V \to A}$ dropped slightly below the prestimulus baseline and returned very slowly to the prestimulus values within the next second.

Even though the temporal development and the frequency spectra were roughly similar in the $nDTF_{A \to V}$ and the $nDTF_{V \to A}$ there were small but important differences. First, there were stimulus-evoked differences in the amplitudes of the $nDTF_{A \to V}$ and the $nDTF_{V \to A}$. After the visual stimulus the nDTF amplitude was significantly higher in the $nDTF_{V \to A}$ than in the $nDTF_{A \to V}$, whereas after the auditory stimulus the $nDTF_{A \to V}$ reached higher values, but only at frequencies above 30 Hz. Second, even though the peaks could be found at all frequency bands in the $nDTF_{V \to A}$ the first peak was strongest at a frequency of 1 Hz and about 32 Hz and the second peak at frequencies of 1 Hz and about 40 Hz. In the $nDTF_{A \to V}$ the highest amplitude values after the first peak could be observed at 1 Hz and at about 35 Hz and after the second peak at 1 Hz and about 45 Hz.

3.1.2   Animals Receiving First the Tone, Then the Light Stimulus (AV-Animals)

In animals receiving first the light stimulus and then the tone stimulus, three positive peaks developed after stimulation. As in the VA-animals the $nDTF_{A \to V}$ and $nDTF_{V \to A}$ were similar to each other (figure 1, left). The first peak could be found between the tone and the light stimulus, at about -40 ms. The second and the third peak occurred after the light stimulus at about 170 ms and 330 ms. And as in the VA-animals in the AV- animals after the auditory stimulus (here the first stimulus) the amplitude of the $nDTF_{A \to V}$ significantly exceeded the amplitude of the $nDTF_{V \to A}$ for frequencies above 20 Hz, whereas after the visual stimulus amplitudes were significantly higher in the $nDTF_{V \to A}$ (figure 1C, left). Thus, the sign of the difference between the $nDTF_{A \to V}$ and the $nDTF_{V \to A}$ depended on the type of the stimulus (auditory or visual) and not on the order of stimulus presentation.

The peaks ran through all frequencies from 0 to 100 Hz. The first peak of the $nDTF_{A \to V}$ was most pronounced at 1 Hz and at about 42 Hz, the second peak at 1 Hz, at about 32 Hz and at 100 Hz. The first peak of the $nDTF_{V \to A}$ reached highest values at 1 Hz and at 35 Hz, the second peak had its highest amplitude at 1 Hz and at 28 Hz. For the third peak the amplitude was most prominent at 1 Hz.

3.2 Development of the Amplitude of $nDTF_{A \to V}$ and $nDTF_{V \to A}$ within the Sessions

To investigate the development of the effects within the sessions we divided the 750 trials of each session into windows of 125 trials from the start to the end of each session. Averaging was done across the trials of each trial window, but separately for the time windows within the course of each trial. Trials from all sessions were included in the average. As for the majority of the animals the nDTF-amplitude increased or decreased fairly smoothly within the sessions we decided to characterize the effects by linear regression analysis. The slope of the regression line fitted to the observed data points was subjected to statistical testing using the bootstrap technique (for details see Methods).

3.2.1 VA-Animals

In figure 3A and 3B the development of the nDTF amplitude of the first and the second peak within the sessions is depicted, averaged across all four animals which received the light stimulus first. Roughly, most of the effects could be observed over the whole range of frequencies tested, in figure 3 we selected nDTF peaks at a frequency of 40 Hz for illustration. Nevertheless, effects did not always reach significance at all frequencies tested, see table 1 and table 2 for more detailed information on the development of peaks at other frequencies).

After the first (visual) stimulus the amplitude of the first peak increased in the $nDTF_{A \to V}$ and decreased in the $nDTF_{V \to A}$ (figure 2A, left). At the beginning of the session the amplitude was higher in the $nDTF_{V \to A}$ than in the $nDTF_{A \to V}$, thus the amplitude difference between the $nDTF_{A \to V}$ and the $nDTF_{V \to A}$ decreased significantly over the session (figure 2A, right).

After the second (auditory) stimulus the amplitude of the second peak increased both in the $nDTF_{A \to V}$ and the $nDTF_{V \to A}$ (figure 2B, left). Importantly, the increase of the $nDTF_{A \to V}$ exceeded the increase of the $nDTF_{V \to A}$, gradually increasing the difference between the $nDTF_{A \to V}$ and the $nDTF_{V \to A}$ (figure 2B, right).

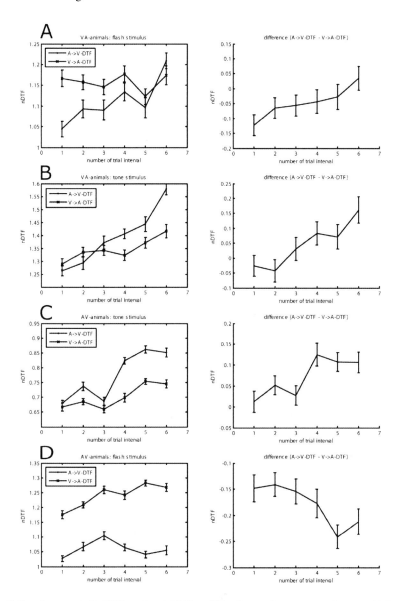

Fig. 2 Development of nDTF-peaks at 40 Hz within the sessions averaged across trial windows of 125 trials stepped at intervals of 125 trials through all sessions. AB: animals receiving first the light. CD: animals receiving first the tone. Left: AC: Development of the average amplitude peak after the first stimulus in the nDTF$_{A\rightarrow V}$ and nDTF$_{V\rightarrow A}$. BD: Development of the average amplitude peak after the second stimulus in the nDTF$_{A\rightarrow V}$ and nDTF$_{V\rightarrow A}$. Right: Amplitude of the nDTF$_{V\rightarrow A}$ peak subtracted from the amplitude of the nDTF$_{A\rightarrow V}$ peak shown in the left figures. Error bars denote the standard error of the mean, averaged across animals.

3.2.2 AV-Animals

Similar to the nDTF development in VA-animals after the second (auditory) stimulus, in the AV-animals after the first (auditory) stimulus the amplitude increased both in the $nDTF_{A \to V}$ and the $nDTF_{V \to A}$ (figure 2C, left). The increase was more pronounced in $nDTF_{A \to V}$, further increasing the difference between the $nDTF_{A \to V}$ and the $nDTF_{V \to A}$ (figure 2C, right).

Interestingly, after the second (visual) stimulus, the behaviour of the nDTF in the AV-animals did not resemble the behaviour of the nDTF after the first (visual) stimulus in the VA-animals. In the AV-animals the amplitude of the $nDTF_{V \to A}$ increased after the visual stimulus, the amplitude of the $nDTF_{A \to V}$ decreased slightly in some animals, in other animals an increase could be observed (see figure 2D left and table 1). As after the visual stimulus the amplitude of the $nDTF_{V \to A}$ was higher than the amplitude of the $nDTF_{A \to V}$ already at the beginning of the sessions, the difference between the $nDTF_{A \to V}$ and the $nDTF_{V \to A}$ further increased during the course of the sessions (figure 2D, right).

3.3 Trial-to-Trial Variability of the Stimulus-Induced Changes in the nDTF

The short-time nDTF of single-trials appeared to be highly unpredictable in amplitude. In general, both the $nDTF_{A \to V}$ and the $nDTF_{V \to A}$ displayed one or two peaks with the frequency of the peak varying strongly across the whole range of frequencies measured. To characterize the trial-to-trial variability the rates of peak occurrences at different frequencies were determined separately for each time point of the trial. In Figure 3 we show the data of a VA-animal. To illustrate the effect of the stimulation, we subtracted the rates of peak occurrences averaged across the prestimulus interval from the rates of peak occurrences of the poststimulus interval. It can be readily observed that the shapes of the poststimulus distributions of the $nDTF_{A \to V}$ (figure 3A) and the $nDTF_{V \to A}$ (figure 3B) were almost identical. Moreover, both distributions were highly similar in shape to the all-trial average of the nDTF of the VA-animals (figure 1): the amplitudes of the all-trial average were high at frequencies we also observed high rates of peak occurrences.

In the all-trial average we observed clear differences between the amplitudes of the $nDTF_{A \to V}$ and the $nDTF_{V \to A}$ in that the AV-DTF attained higher values after the auditory stimulus and the $nDTF_{V \to A}$ attained higher values after the visual stimulus. Interestingly, similar differences could not be observed in the rate distributions (figure 3C). The difference between the two distributions were almost zero at most frequencies and time points.

In figure 3(D and E) we plotted the averaged *amplitudes* of the peaks at different frequencies, again separately for the different time points of the trial. Also the amplitudes of the peaks were highly similar in shape to the amplitudes of the all-trial average. Moreover, in contrast to the rate distributions, for the distributions of peak amplitudes we now observed the differences between the peak amplitudes of

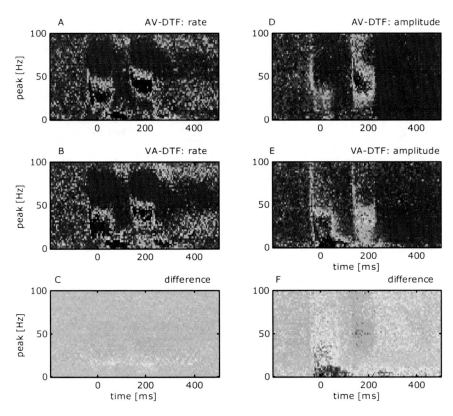

Fig. 3 A & B: Rates of occurrence of peaks in the single-trial $nDTF_{A \to V}$ (A) and $nDTF_{V \to A}$ (B) at different time points after stimulation. C: rates of peak occurrences of the $nDTF_{A \to V}$ subtracted from rates of peak occurrences of the $nDTF_{V \to A}$. D & E: Amplitudes of peaks in the single-trial $nDTF_{A \to V}$ (D) and $nDTF_{V \to A}$ (E) at different time points after stimulation. F: Amplitudes of peaks of the $nDTF_{A \to V}$ subtracted from amplitudes of peaks of the $nDTF_{V \to A}$. Data from an animal receiving first the light.

the $nDTF_{A \to V}$ and $nDTF_{V \to A}$ (figure 3F) corresponding closely to those found in the all-trial average: after the auditory stimulus the amplitudes of the peaks of the $nDTF_{A \to V}$ exceeded the amplitudes of the peaks of the $nDTF_{V \to A}$, after the visual stimulus the peak amplitudes of the $nDTF_{V \to A}$ attained higher values than those of the $nDTF_{A \to V}$.

To sum up, the single-trial dynamics characterized by the rates of peak occurrences appeared to correspond closely in the $nDTF_{A \to V}$ and the $nDTF_{V \to A}$, suggesting that a common waveform occurred in both nDTFs in the single trials. However, within the single trials on average the nDTF differed in amplitude, suggesting directive influences between the cortices (see Discussion for the interpretation of nDTF amplitudes).

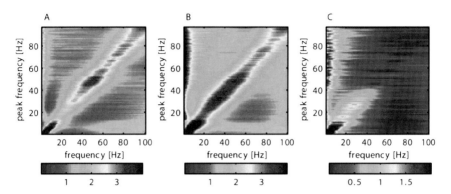

Fig. 4 Single-trial nDTF$_{A \to V}$ selectively averaged for the frequency of one of its peaks. A: data from the prestimulus interval. B: data 0 ms to 200 ms after the auditory stimulus. C. Differences of the averages of the pre- and poststimulus interval.

When looking only at peak amplitudes, it seems that some of the information contained in the nDTF was lost. However, as the peak was largest it also had the strongest influence on the all-trial average of the nDTF. In figure 4 we averaged the single-trial nDTFs selectively for one of its peaks, both for the pre- and the poststimulus interval. In general the selective averages were largest at its peaks. Interestingly, for nDTF-averages with peaks below 50 Hz the stimulus induced effects was most pronounced around the peak frequency used for averaging. This finding illustrates the clear differences in the evoked responses observed in single-trials.

Next we addressed the question whether the single-trial dynamics would change within a session. In figure 5 we presented rates of peak occurrences from the first 250 trial (figure 5, left) and the last 250 trials (figure 5, middle) of the sessions, averaged for animals receiving first the light and then the tone stimulus. It can be seen that from the beginning to the end of the sessions there were no strong changes in the rate of peak occurrences. In contrast, the amplitudes of the peaks determined from the first 250 trial and the last 250 trials of the session, exhibited changes in amplitude similar to those observed in the all-trial average.

3.4 Development of the Amplitude nDTF$_{A \to V}$ and nDTF$_{V \to A}$ across the Sessions

To examine effects of long-term adaptation the nDTF amplitude of the first 100 trials was averaged separately for each session. The development of the amplitude averages across sessions was examined by linear regression analysis and the significance of the slope was tested using the bootstrap technique. In the following, effects are reported for a chosen significance level of 0.05.

Even though some significant trends could be observed, results were not consistent among animals. In the VA-animals in one animal a decrease could be observed in the amplitude of the nDTF$_{A \to V}$ at the beginning of the first stimulus, but

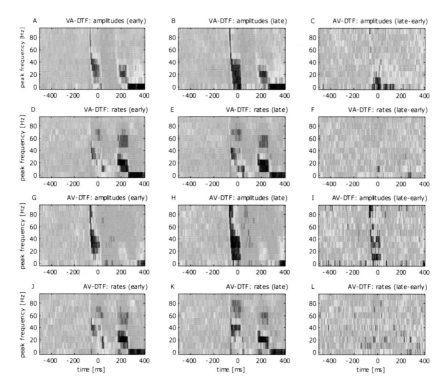

Fig. 5 Development of the rates of occurrence (D,E,F,J,K,L) and the amplitudes (A,B,C,G,H,I) of peaks of single-trial nDTFs from the first 250 trials of sessions (A,B,D,G: early) to the last 250 trials of the sessions (B,E,H,I: late). C,F,I,L: difference between early and late trials of the sessions. A,B,C,D,E,F: data from nDTFVA. GHIJKL: data from nDTFAV. Data from all animals receiving first the tone.

an increase could be found only 20 ms after the beginning of the first stimulus. In a second animal there was an increase in the amplitude of the $nDTF_{A \to V}$ after the second stimulus. In the amplitude of the $nDTF_{V \to A}$ of two VA-animals decreases could be observed after the first and after the second stimulus, in a third animal an increase was found after the second stimulus. All these results could be observed for the majority of examined frequencies.

In the $nDTF_{A \to V}$ of the AV-animals at many frequencies no clear developmental trend could be observed, but at frequencies below 10 Hz in two animals there was an increase in amplitude both after the first and the second stimulus, whereas in one animal a decrease could be found after both stimuli. In the amplitude of the $nDTF_{V \to A}$ increases could be observed at various frequencies and time points after stimulation.

4 Discussion

When pairs of auditory and visual stimuli were presented repeatedly at a constant stimulus-onset-asynchrony, within the adaptation sessions we observed modulations of the evoked amplitude of the Directed Transfer Function with increasing number of stimulus repetitions. This finding suggests that the modes of interaction between the primary auditory and primary visual cortex adapted to the prolonged exposure with asynchronous pairs of auditory and visual stimuli. Across the adaptation sessions, no coherent development could be observed indicating that there were no long-term effects on the cross-cortical interactions. In the following we discuss possible processes evoked by the repeated asynchronous presentation of audiovisual stimuli and their possible effects on the amplitude of the nDTF. To prepare the discussion some general considerations with respect to the interpretation of nDTF amplitudes seem appropriate.

4.1 Interpretation of DTF Amplitudes

Long-range interaction processes have been frequently associated with coherent oscillatory activity between cortical areas [10,73,72,85]. Moreover, it has been shown that the oscillatory activity in one cortical area can be predicted by earlier measurement of another cortical area using the DTF [43,44,50,51,30,58,54], indicating that the oscillatory activity might signal directional influences between the cortices.

In our data we observed that independent of the stimulus order, when an auditory stimulus was presented, the amplitude of the $nDTF_{A \to V}$ exceeded the amplitude of the $nDTF_{V \to A}$, whereas after the visual stimulus the amplitude of the $nDTF_{V \to A}$ reached higher values. Given these observation, one might feel tempted to conclude that after the auditory stimulus, the auditory cortex sent information about the auditory stimulus to the visual cortex and after the visual stimulus, the visual cortex informed the auditory cortex about the visual stimulus. However, as [17] demonstrated in a series of simulation studies, there is no straightforward way to infer directed crosscortical interactions from the information provided by the DTF. Specifically, from DTF amplitudes alone it is not possible to tell whether the information flow is unidirectional, bidirectional or even multidirectional including additional brain areas.

For example, let us consider the situation after the presentation of the auditory stimulus when the amplitude of the $nDTF_{A \to V}$ attained higher values than the amplitude of the $nDTF_{V \to A}$. Firstly, this result might indicate that there was an unidirectional influence from the auditory to the visual cortex with the size of the amplitude difference positively correlating with the delay in the information transfer. Secondly, this finding could also reflect a reciprocal influence between the auditory and visual cortex, but with the influence from auditory cortex either larger in amplitude or lagged relative to the influence from the visual cortex. Thirdly, additional unobserved structures might be involved sending input slightly earlier to the auditory cortex than to the visual cortex.

4.2 The Development of the nDTF Amplitude within Sessions

The development of the nDTF after the *auditory* stimulus did not seem to depend strongly on the order of stimulus presentation. Independent of whether an auditory or a visual stimulus was presented first, after the auditory stimulus the peak amplitude of both the $nDTF_{A \to V}$ and $nDTF_{V \to A}$ increased. Noteworthy, the increase was more pronounced in the $nDTF_{A \to V}$ than in the $nDTF_{V \to A}$ further increasing the difference between the amplitudes of the $nDTF_{A \to V}$ and the $nDTF_{V \to A}$. Using the interpretation scheme introduced above, under the assumption of unidirectional interaction, the influence from the auditory to the visual cortex not only increased in strength but also the lag with which the input is sent became larger with increasing number of stimulus repetitions? In case of bidirectional interaction, influences from both sides increased, but the influence form the auditory cortex became stronger relatively to the influence from the visual cortex. Last, in case of multidirectional interaction the influence of a third structure to both the auditory and the visual cortex might become more pronounced, but at the same time the temporal delay input is sent to the visual cortex relatively to the delay input is send to the auditory cortex is increased even further. All these three interpretations have in common that the interaction did not only gather in strength, but also the mode of the interaction changed.

In contrast to the development of the nDTF after the auditory stimulus the development of the nDTF after the visual stimulus clearly depended on the order of stimulus presentation. When the visual stimulus was presented first, contrary to expectations, the amplitude of the $nDTF_{V \to A}$ decreased with increasing number of stimulus repetitions, whereas the amplitude of the $nDTF_{A \to V}$ increased in the majority of the animals. Thus, assuming that unidirectional influence underlies our data, this finding might reflect that the visual cortex sends influences to the auditory cortex at increasingly shorter delays. In case of bidirectional interaction the input from the visual cortex decreases whereas the input from the auditory cortex increases. Last, under assumption of multidirectional interaction a hypothetical third structure might still send its input earlier to the visual cortex, but the delay became diminished with increasing number of stimulus repetitions.

When the visual stimulus was presented as the second stimulus the behavior of the nDTF showed some resemblance to the behaviour of the nDTF after the auditory stimulus. More precisely, both the peak amplitude of the $nDTF_{A \to V}$ and the $nDTF_{V \to A}$ increased within the sessions. But importantly, now the increase was stronger in the $nDTF_{V \to A}$.

4.3 Single-Trial Variability of the nDTF

The amplitude of the nDTF exhibited a strong trial-to-trial variability: it usually displayed one or two peaks with the frequency of the peaks differing strongly from trial to trial. The observation that the modes of crosscortical interactions might be highly variable has also been reported in other studies on the dynamics of crosscortical interactions [13].

The modulations in the rates of peak occurrence observed after stimulation strongly agreed with the changes in the amplitudes observed in the all-trial average: high rates of peak occurrences could be observed at frequencies we also found high amplitudes. When the single-trial nDTFs were averaged selectively in dependence of the frequency of one of their peaks, the evoked response observed in the selective averages deviated from the evoked response found in the all-trial average. Thus we did not find strong evidence for a fixed stimulus-evoked response, constantly occurring in all trials independent of the state of the system. Instead, the trial-to-trial variability did not average out by averaging across trials but appeared to have a strong influence on the shape of the all-trial average.

The evoked changes in the distributions of peaks occurrences were almost identical in the $nDTF_{A \rightarrow V}$ and the $nDTF_{V \rightarrow A}$, indicating that the $nDTF_{A \rightarrow V}$ and $nDTF_{V \rightarrow A}$ shared highly similar waveforms in the single-trials and did not differ much in their single-trial dynamics. However, the $nDTF_{A \rightarrow V}$ - and $nDTF_{V \rightarrow A}$ clearly differed in the relative amplitudes of their waveforms indicating the presence of directive interactions between the cortices.

With increasing number of repetitions of the asynchronous stimuli no substantial changes in the rates of peak occurrences were discernable. This observation suggests that the prolonged exposure to the asynchronous stimulation did not affect the constant waxing and waning of spectral activities observed in the ongoing activity. With other words, the amplitude adaptations observed in the all-trial average appeared to be due to changes in the amplitudes of the induced responses within the trials rather than to their trial-to-trial variability.

4.4 Hypothetical Processes Underlying the Observed Changes in the nDTF Amplitudes

As paired-stimulus adaptation protocols, similar to the one used in the present study, have been shown to induce recalibration of temporal-order judgment in humans (e.g. [33,94], some of the described effects on the directed information transfer could possibly underlie such recalibration functions. For example, the characteristic developmental trend after the second stimulus was an increase in both $nDTF_{A \rightarrow V}$ and $nDTF_{V \rightarrow A}$ with the increase stronger in the nDTF sending information from the structure the stimulus had been presented to, namely in the $nDTF_{V \rightarrow A}$ after the visual stimulus, and in the $nDTF_{A \rightarrow V}$ after the auditory stimulus. One might hypothesize that the increase in the interactions between the auditory and the visual cortex after the second stimulus mirrored the integration of the auditory and the visual information resulting in a reduction of their perceived temporal distance.

However, we have to take into account that behavioral studies on recalibration of temporal-order judgment typically demonstrated a shift of the entire psychometric function (i.e. at many stimulus onset synchronies and irrespective of the stimulus order). This finding is remarkable given that the recalibration was induced by presentation of stimuli at a constant order and stimulus-onset-asynchrony. In our data the behavior of the nDTF after the visual stimulus clearly depended on the stimulus order, thus this findings disagrees with the results of the behavioral recalibration experiments.

The independence of the recalibration effect on the stimulus order implies that during the recalibration process the temporal perception of multimodal stimuli is not recalibrated *relative* to each other but perception is simply speeded up or slowed down in one modality.

In our data we did not find any indications for an increase in the speed of stimulus processing in form of a change in the latencies of the nDTF peaks. However, it also might appear surprising if the temporal compensation mechanisms were so simple that they might be readily observed in a decrease in the latency of crosscortical interactions.

To decide whether the changes in the nDTF we observed were neural correlates of the recalibration of temporal perception the repetition of our experiment in combination with a behavioral test is necessary.

Many other cognitive processes might have been evoked by the paired presentation of stimuli. For example, in accordance to the unity assumption (e.g. [7,96,95]Bedford, 2001; two stimuli from different sensory modalities will be more likely regarded as deriving from the same event when they are presented in close temporal congruence. The increase in the amplitude of the nDTF after the second stimulus might indicate the binding of the stimuli into a coherent perception. Changes in the nDTF-amplitude before the second stimulus might indicate the expectation of the second stimulus. Several other studies have demonstrated increases in coherent activity associated with anticipatory processing (e.g. [74,93,32,55]. To clarify whether the observed changes might have something to do with stimulus association or expectation processes the repetition of this experiment with anesthetized animals might be helpful.

As we presented our stimuli with constant audiovisual lag, also mechanisms of lag detection could have been evoked. There are first studies on the neural correlates of synchronous and asynchronous stimulus presentation ([59,15,77]. [77] could observe changes in the oscillatory activity with synchronous or asynchronous presentation of audiovisual stimuli, suggesting that the changes in the oscillatory interaction between cortices we observed in the nDTF might indicate the alerting of lag detectors.

References

1. Akaike, H.: A new look at statistical model identification. Trans. Aut. Contr. 19, 716–723 (1974)
2. Alais, D., Carlile, S.: Synchronizing to real events: Subjective audiovisual alignment scales with perceived auditory depth and speed of sound. Proc. Nat. Acad. Sci. USA 102(6), 2244–2247 (2005)
3. Arnold, D.H., Johnston, A., Nishida, S.: Timing sight and sound. Vis. Res. 45, 1275–1284 (2005)
4. Astolfi, L., Cincotti, F., Mattia, D., Marciani, M.G., Baccala, L.A., de Vico Fallani, F., Salinari, S., Ursino, M., Zagaglia, M., Ding, L., Edgar, J.C., Miller, G.A., He, B., Babiloni, F.: Comparison of different cortical connectivity estimators for high-tech resolution EEG Recordings. Hum. Brain Map 28, 143–157 (2007)

5. Baylor, D.A., Nunn, B.J., Schnapf, J.L.: The photocurrent, noise and spectral sensitivity of rods of the monkey Macaca fascicularis. J. Physiol. 357, 575–607 (1984)
6. Baylor, D.A., Nunn, B.J., Schnapf, J.L.: Spectral sensitivity of cones of the monkey Macaca fascicularis. J. of Physiol. 390, 124–160 (1987)
7. Bedford, F.L.: Toward a general law of numerical/object identity. Cur. Psych. Cog. 20(3-4), 113–175 (2001)
8. Bertelson, P., de Gelder, B.: The psychology of multimodal perception. In: Spence, C., Driver, J. (eds.) Crossmodal space and crossmodal attention, pp. 141–177. Oxford University Press, Oxford (2004)
9. Bizley, J.K., Nodal, F.R., Bajo, V.M., Nelken, I., King, A.J.: Physiological and anatomical evidence for multisensory interactions in auditory cortex. Cereb. Cortex 17, 2172–2198 (2007)
10. Bressler, S.L., Coppola, R., Nakamura, R.: Episodic multiregional cortical coherence at multiple frequencies during visual task performance. Nature 366, 153–156 (1993)
11. Bressler, S.L.: Large scale cortical networks and cognition. Brain Res. Rev. 20, 288–304 (1995)
12. Bressler, S.L.: Interareal synchronization in the visual cortex. Behav. Brain Res. 76, 37–49 (1996)
13. Bressler, S.L., Kelso, J.S.A.: Cortical coordination dynamics and cognition. Trends Cog. Sci. 5, 26–36 (2001)
14. Brosch, M., Selezneva, E., Scheich, H.: Nonauditory events of a behavioral procedure activate auditory cortex of highly trained monkeys. J. Neurosci. 25(29), 6796–6806 (2005)
15. Bushara, K.O., Grafman, J., Hallet, M.: Neural correlates of audio-visual stimulus onset asynchrony detection. J. Neurosci. 21(1), 300–304 (2001)
16. Cahill, L., Ohl, F.W., Scheich, H.: Alternation of auditory cortex activity with a visual stimulus through conditioning: a 2-deoxyglucose analysis. Neurobiol. Learning and Memory 65(3), 213–222 (1996)
17. Cassidy, M., Brown, P.: Spectral phase estimates in the setting of multidirectional coupling. J. Neurosci. Meth. 127, 95–103 (2003)
18. Cobbs, E.H., Pugh Jr., E.N.: Kinetics and components of the flash photocurrent of isolated retinal rods of the larval salamander, Ambystoma tigrinum. J. Physiol. 394, 529–572 (1987)
19. Corey, D.P., Hudspeth, A.J.: Response latency of vertebrate hair cells. Biophys. J. 26, 499–506 (1979)
20. Corey, D.P., Hudspeth, A.J.: Analysis of the microphonic potential of the bullfrog's sacculus. J. Neurosci. 3, 942–961 (1983)
21. Crawford, A.C., Evans, M.G., Fettiplace, R.: The actions of calcium on the mechanoelectrical transducer current of turtle hair cells. J. Physiol. 491, 405–434 (1991)
22. Crawford, A.C., Fettiplace, R.: The mechanical properties of ciliary bundles of turtle cochlear hair cells. J. Physiol. 364, 359–379 (1985)
23. DeGuzman, G.C., Kelso, J.A.S.: Multifrequency behavioral patterns and the phase attractive circle map. Biol. Cybern. 64, 485–495 (1991)
24. Ding, M., Bressler, S.L., Yang, W., Liang, H.: Short-window spectral analysis of cortical event-related potentials by adaptive autoregressive modelling: data preprocessing, model validation, variability assessment. Biol. Cybern. 83, 35–45 (2000)
25. Eichler, M.: On the evaluation of information flow in multivariate systems by the directed transfer function. Biol. Cybern. 94, 469–482 (2006)

26. Engel, G.R., Dougherty, W.G.: Visual-auditory distance constancy. Nature 234, 308 (1971)
27. Efron, B., Tibshirani, R.J.: An introduction to the bootstrap. Chapman & Hall/CRC, Boca Raton, Florida (1993)
28. Fain, G.L.: Sensory transduction. Sinauer Associates, Sunderland (2003)
29. Friston, K.J.: The labile brain. I. Neuronal transients and nonlinear coupling. Phil. Trans. R Soc. Lond. B 355, 215–236 (2000)
30. Franaszczuk, P.J., Bergey, G.K.: Application of the directed transfer function method to mesial and lateral onest temporal lobe seizures. Brain Topography 11, 13–21 (1998)
31. Freeman, W.J.: Neurodynamics: An exploration in mesoscopic brain dynamics. Springer, Heidelberg (2000)
32. Fries, P., Reynolds, J.H., Rorie, A.E., Desimone, R.: Modulation of oscillatory neuronal synchronization by selective visual attention. Sci. 291, 1560–1563 (2001)
33. Fujisaki, W., Shinsuke, S., Makio, K., Nishida, S.: Recalibration of audiovisual simultaneity. Nature Neuroscience 7, 773 (2004)
34. Fujisaki, W., Nishida, S.: Temporal frequency characteristics of synchrony-asynchrony discrimination of audio-visual signals. Exp. Brain Res. 166, 455–464 (2005)
35. Fujisaki, W., Koene, A., Arnold, D., Johnston, A., Nishida, S.: Visual search for a target changing in synchrony with an auditory signal. Proc. R Soc. B 273, 865–874 (2006)
36. Fujisaki, W., Nishida, S.: Top-down feature based selection of matching feature for audio-visual synchrony discrimination. Neurosci. Let. 433, 225–230 (2008)
37. Harrar, V., Harris, L.R.: Simultaneity constancy: Detecting events with touch and vision. Exp. Brain Res. 166, 465–473 (2005)
38. Harrar, V., Harris, L.R.: The effects of exposure to asynchronous audio, visual, and tactile stimulus combination on the perception of simultaneity. Experimental Brain Research 186, 517–524 (2008)
39. Hanson, J.V.M., Heron, J., Whitaker, D.: Recalibration of perceived time across sensory modalities. Exp. Brain Res. 185, 347–352 (2008)
40. Heron, J., Whitaker, D., McGraw, P., Horoshenkov, K.V.: Adaptation minimizes distance-related audiovisual delays. J. Vis. 7(13), 5:1–5:8 (2007)
41. Hestrin, S., Korenbrot, J.I.: Activation kinetics of retinal cones and rods: response to intense flashes of light. J. Neurosci. 10, 1967–1973 (1990)
42. Kaminski, M., Blinowska, K.J.: A new method for the description of the information flow in the brain structures. Biol. Cybern. 65, 203–210 (1991)
43. Kaminski, Blinowska, K.J., Szelenberger, W.: Topographic analysis of coherence and propagation of EEG activity during sleep and wakefulness. Electroencephal. Clin. Neuro-physiol. 102, 216–277 (1997)
44. Kaminski, M., Ding, M., Trucculo, W.A., Bressler, S.L.: Evaluating causal relations in neural systems: Granger causality, directed transfer function and statistical assessment of significance. Biol. Cybern. 85, 145–157 (2001)
45. Kay, M.S.: Modern spectral estimation. Prentice-Hall, Englewood Cliffs (1987)
46. Kayser, C., Petkov, C., Logothetis, N.K.: Visual modulation of neurons in auditory cortex. Cerebral Cortex 18, 1560–1574 (2008)
47. Keetels, M., Vroomen, J.: No effect of auditory-visual spatial disparity on temporal recalibration. Exp. Brain Res. 182, 559–565 (2007)
48. King, A.J., Palmer, A.R.: Integration of visual and auditory information in bimodal neurons in the guinea-pig superior colliculus. Exp. Brain Res. 60, 492–500 (1985)

49. Kopinska, A., Harris, L.R.: Simultaneity constancy. Perception 33, 1049–1060 (2004)
50. Korzeniewska, A., Kasicki, S., Kaminski, M., Blinowska, K.J.: Information flow between hippocampus and related structures during various types of rat's behavior (1997)
51. Korzeniewska, A., Manczak, M., Kaminski, M., Blinowska, K., Kasicki, S.: Determination of information flow direction among brain structures by a modified directed transfer function (dDTF) method. J. Neurosci. Meth. 125, 195–207 (2003)
52. Kus, R., Kaminski, M., Blinowska, K.J.: Determination of EEG activity propagation: pairwise versus multichannel estimate. IEEE Trans. Biomed. Eng. 51, 1501–1510 (2004)
53. Lewald, J., Guski, R.: Auditory-visual temporal integration as a function of distance: no compensation for sound-transmission time in human perception. Neurosci. Let. 357, 119–122 (2004)
54. Liang, H., Ding, M., Nakamura, R., Bressler, S.L.: Causal influences in primate cerebral cortex. Neuroreport 11(13), 2875–2880 (2000)
55. Liang, H., Bressler, S.L., Ding, M., Truccolo, W.A., Nakamura, R.: Synchronized activity in prefrontal cortex during anticipation of visuomotor processing. Neuroreport 13(16), 2011–2015 (2002)
56. Lütkepohl, H.: Indroduction to multiple time series analysis, 2nd edn. Springer, Heidelberg (1993)
57. Marple, S.L.: Digital spectral analysis with applications. Prentice-Hall, Englewood Cliffs (1987)
58. Medvedev, A., Willoughby, J.O.: Autoregressive modeling of the EEG in systemic kainic acid-induced epileptogenesis. Int. J. Neurosci. 97, 149–167 (1999)
59. Meredith, M.A., Nemitz, J.W., Stein, B.E.: Determinants of multisensory integration in Superior Colliculus Neurons I. Temporal factors. J. Neurosci. 7(10), 3212–3229 (1987)
60. Miyazaki, M., Yamamoto, S., Uchida, S., Kitazawa, S.: Baysian calibration of simultaneity in tactile temporal order judgement. Nat. Neurosci. 9, 875–877 (2006)
61. Morein-Zamir, S., Soto-Faraco, S., Kingstone, A.: Auditory capture of vision: Examining temporal ventriloquism. Brain Research and Cognitive Brain Research 17, 154–163 (2003)
62. Musacchia, G., Schroeder, C.E.: Neural mechanisms, response dynamics and perceptual functions of multisensory interactions in auditory cortex. Hearing Res. 10 (2009) (in press)
63. Navarra, J., Vatakis, A., Zampini, M., Soto-Faraco, S., Humphreys, W., Spence, C.: Exposure to asynchronous audiovisual speech extends the temporal window for audiovisual integration. Cog. Brain Res. 25, 499–507 (2005)
64. Navarra, J., Soto-Faraco, S., Spence, C.: Adaptation to audiovisual asynchrony. Neurosci. Let. 431, 72–76 (2006)
65. Nickalls, R.W.D.: The influences of target angular velocity on visual latency difference determined using the rotating Pulfirch effect. Vis. Res. 36, 2865–2872 (1996)
66. Nuttall, A.H.: Multivariate Linear Predictive Spectral Analysis Employing Weighted Forward and Backward Averaging: A Generalization of Burg's Algorithm, Naval Underwater Systems Center Technical Report 5502, New London (1976)
67. Ohl, F.W., Scheich, H., Freeman, W.J.: Topographic Analysis of Epidural Pure-Tone-Evoked Potentials in Gerbil Auditory Cortex. J. Neurophysiol. 83, 3123–3132 (2000)
68. Ohl, F.W., Scheich, H., Freeman, W.J.: Change in pattern of ongoing cortical activity with auditory learning. Naturet 412, 733–736 (2001)
69. Poppel, E., Schill, K., von Steinbuchel, N.: Sensory integration within temporally neutral systems states: a hypothesis. Naturwissenschaften 77(2), 89–91 (1990)

70. Posner, M.I., Snyder, C.R., Davidson, B.J.: Attention and the detection of signals. J. of Exp. Psychol.: G 109(2), 160–174 (1980)
71. Robson, J.G., Saszik, S.M., Ahmed, J., Frishman, L.J.: Rod and cone contributions to the a-wave of the electroretinogram of the macaque. J. Physiol. 547, 509–530 (2003)
72. Rodriguez, E., et al.: Perception's shadow: long-distance synchronization of neural activity. Nature 397, 430–433 (1999)
73. Roelfsema, P.R., Engel, A.K., König, P., Singer, W.: Visuomotor integration is associated with zero time-lag synchronization among cortical areas. Nature 385, 157–161 (1997)
74. Roelfsema, P.R., Lamme, V.A.F., Spekreijse, H.: Object based attention in the primary auditory cortex of the macaque monkey. Nature 395, 377–381 (1998)
75. Scheider, K.A., Bavelier, D.: Components of visual prior entry. Cog. Psychol. 71, 333–366 (2003)
76. Schlögl, A.: A comparison of multivariate autoregressive estimators. Sig. Proc. 86, 2426–2429 (2006)
77. Senkowski, D., Talsma, D., Grigutsch, M., Herrmann, C.S., Woldorff, M.G.: Good times for multisensory integration: Effects of the precision of temporal synchrony as revealed by gamma band oscillations. Neuropsychologica 45, 561–571 (2007)
78. Stone, J.V.: Where is now? Perception of simultaneity. Proc. R Soc. Lond. B Biol. Sci. 268, 31–38 (2001)
79. Strand, O.N.: Multichannel Complex Maximum Entropy (Autoregressive) Spectral Analysis. IEEE Trans. Autom. Control AC 22, 634–640 (1977)
80. Sugita, Y., Suzuki, Y.: Audiovisual perception. Implicit evaluation of sound arrival time. Nature 421, 911 (2003)
81. Sutton, S., Braren, M., Subin, J., John, E.R.: Evoked Potential Correlates of Stimulus Uncertainty. Science 150, 1178–1188 (1965)
82. Tononi, G., Sporns, O., Edelman, G.M.: A measure for brain complexity: relating functional segregation and integration in the nervous system. Proc. Natl. Acad. Sci. USA 91, 5033–5037 (1994)
83. Tappe, T., Niepel, M., Neumann, O.: A dissociation between reaction time to sinusoidal gratings and temporal order judgement. Perception 23, 335–347 (1994)
84. Titchner, E.B.: Lectures on the elementary psychology of feeling and attention. MacMillian, New York (1908)
85. Varela, F., Lacheaux, J., Rodriguez, E., Martinerie, J.: The brain-web: phase synchronization and large-scale integration. Nat. Rev. Neurosci. 2, 229–239 (2001)
86. Vatakis, A., Navarra, J., Soto-Faraco, S., Spence, C.: Temporal recalibration during asynchronous audiovisual speech research. Exp. Brain Res. 181, 173–181 (2007)
87. Vatakis, A., Navarra, J., Soto-Faraco, S., Spence, C.: Temporal recalibration during asynchronous audiovisual speech perception. Exp. Brain Res. 181, 173–181 (2007)
88. Vatakis, A., Navarra, J., Soto-Faraco, S., Spence, C.: Audiovisual temporal adaptation of speech: temporal order versus simultaneity judgements. Exp. Brain Res. 185, 521–529 (2008)
89. Vatakis, A., Spence, C.: Crossmodal binding: Evaluating the influence of the 'unity assumption' using audiovisual speech stimuli. Percept Psychophys 69(5), 744–756 (2007)
90. Vatakis, A., Spence, C.: Evaluating the influence of the 'unity assumption' on the temporal perception of realistic audiovisual stimuli. Acta Psychol. 127, 12–23 (2008)

91. Virsu, V., Oksanen-Hennah, H., Vedenpää, A., Jaatinen, P., Lahti-Nuuttila, P.: Simultaneity learning in vision, audition, tactile sense and their cross-modal combinations. Exp. Brain Res. 186, 525–537 (2008)
92. Von Békésy, G.: Interaction of paired sensory stimuli and conduction of peripheral nerves. Journal of Applied Physiology 18, 1276–1284 (1963)
93. Von Stein, A., Chiang, C., König, P.: Top-down processing mediated by interarea synchronization. Proceedings of the National Academy of Science US 97, 147148–147153 (2000)
94. Vroomen, J., Keetels, M., de Gelder, B., Bertelson, P.: Recalibration of temporal order perception by exposure to audio-visual asynchrony. Cog. Brain Res. 2, 32–35 (2004)
95. Welch, R.B., Warren, D.H.: Immediate perceptual response to intersensory discrepancy. Psychol. Bulletin 8, 638–667 (1980)
96. Welch, R.B.: Meaning, attention and the unity assumption in the intersensory bias of spatial and temporal perceptions. In: Achersleben, G., Bachmann, T., Müsseler, J. (eds.) Cognitive contributions to the perception of spatial and temporal events, pp. 371–387. Elsevier, Amsterdam (1999)
97. Wilson, J.A., Anstis, S.M.: Visual delay as a function of luminance. The Am. J. Psycho. 82, 350–358 (1996)
98. Zampini, M., Shore, D.I., Spence, C.: Audiovisual prior entry. Neurosci. Let. 381, 217–222 (2005)

Discrimination of Locomotion Direction at Different Speeds: A Comparison between Macaque Monkeys and Algorithms

Fabian Nater*, Joris Vangeneugden*, Helmut Grabner,
Luc Van Gool, and Rufin Vogels

Abstract. Models for visual motion perception exist since some time in neurophysiology as well as computer vision. In this paper, we present a comparison between a behavioral study performed with macaque monkeys and the output of a computational model. The tasks include the discrimination between left and right walking directions and forward *vs.* backward walking. The goal is to measure generalization performance over different walking and running speeds. We show in which cases the results match, and discuss and interpret differences.

1 Introduction

Correctly recognizing biological motion is of utmost importance for the survival of all animals. The computer vision field has a long tradition in modeling human motion, especially of walking and running. Of those models, some recent ones have been inspired by neurophysiology. Human locomotion consists of motion patterns that involve different movements of all limbs and are therefore widely used to study visual motion perception on a psychophysical level but also for modeling with computational algorithms. Studies on action recognition in both fields suggest that human actions can be described using appearance/form or motion cues (*e.g.*, [1], [2]).

Fabian Nater · Helmut Grabner · Luc Van Gool
Computer Vision Laboratory, ETH Zurich, Switzerland

Joris Vangeneugden · Rufin Vogels
Laboratorium voor Neuro- en Psychofysiologie, K.U. Leuven, Belgium

Luc Van Gool
ESAT - PSI / IBBT, K.U. Leuven, Belgium

* Equally contributing first authors.
  Work is funded by the EU Integrated Project DIRAC (IST-027787).

D. Weinshall, J. Anemüller, and L. van Gool (Eds.): DIRAC, SCI 384, pp. 181–190.
springerlink.com   © Springer-Verlag Berlin Heidelberg 2012

In this work, we set out to specifically investigate the perception of loco-motion direction. While discrimination between right- and leftward walking is possible based on shape cues only (e.g. comparing momentary body poses), discriminating between forward and backward walking at least requires mo-tion for a successful distinction [3]. A recent behavioral study in macaque monkeys investigated the perception of walking direction and how well these animals generalized from a trained categorization of walking to other walking speeds and running [4]. The question now arises how state-of-the-art meth-ods in computer vision relate to these findings. More specifically, we focus on a recently proposed approach which encodes typical appearance and mo-tion patterns in a hierarchical framework [5]. The two hierarchies of this model are intended to mirror the subdivision into "snapshot" and "motion" sensitive neurons that have been found in the brain of the macaque monkey, more specifically in the superior temporal sulcus within the temporal lobe [6]. While "snapshot" neurons encode for static body poses, "motion" neurons are driven by movement (i.e. kinematics). In the present paper we aim to compare the performances of this algorithm and of behavioral macaque re-sponses regarding the visual coding of human locomotion at different walking and running speeds.

2 Behavioral Study

2.1 Subjects and Apparatus

Three rhesus monkeys (*Macaca mulatta*) served as subjects in this study. The heads of the monkeys were kept immobilized during the sessions (approx. 3h/day) in order to capture the position of one eye via an infrared tracking device (EyeLink II, SR Research, sampling rate 1000 Hz). Eye positions were sampled to assure that the subjects fixated the stimuli. In order to obtain a juice reward (operant conditioning), successful fixation, within a predefined window measuring $1.3° - 1.7°$, and a correct direct saccade towards one of the response targets were required, see Fig. 1.

Each trial started with the presentation of a small red square at the center of the screen ($0.12°$ by $0.12°$). The subjects had to fixate this square for 500 ms, followed by the presentation of the stimulus (duration $= 1086$ ms; 65 frames at a 60 Hz frame rate). Before making a direct eye movement saccade to one of the two response targets, the monkeys had to fixate the small red square for another 100 ms. During the complete trial duration, monkeys had to maintain their eye position within the predefined window. Failure to do so resulted in a trial abort. Response targets were located at $8.4°$ eccentricity, either on the right, left of upper part of the screen. The stimuli, described below, measured approximately $6°$ by $2.8°$ degrees (height/ width at the maximal lateral extension respectively).

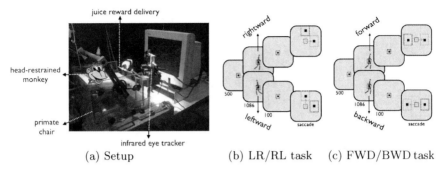

(a) Setup   (b) LR/RL task   (c) FWD/BWD task

Fig. 1 Experimental setup for the behavioral study (a), illustration of the performed tasks (b,c). The gray fields indicate what was presented to the monkeys on the screen, with the respective durations indicated below each screenshot (times expressed in ms). The dotted rectangles around the dark fixation point and the response targets represent the windows in which eye positions had to remain prior to, during and after stimulus presentation (former case) or eye movements had to land, indicating the monkeys' decision (the latter case). Highlighted in light grey are the correct targets the monkeys had to saccade to in order to obtain a juice reward.

All animal care, experimental and surgical protocols complied with national and European guidelines and were approved by the K.U. Leuven Ethical Committee for animal experiments.

2.2 Stimuli

Stimuli were generated by motion-capturing a male human adult of average physical constitution walking at 2.5, 4.2 or 6 or running at 8, 10 or 12 km/h. Specifications of the procedure can be found in [4]. We rendered enriched stimulus versions by connecting the joints by cylinder-like primitives, yielding *humanoid* renderings. Importantly, all stimuli were displayed resembling treadmill locomotion, *i.e.*, devoid of any extrinsic/translatory motion component, see Fig. 2.

2.3 Tasks and Training

The three monkeys were extensively trained in discriminating between different locomotion categories. In a first task, they were instructed to discriminate between different facing directions (LR/RL task) when observing the stimulus (video) that shows a person that is either walking towards the right (LR_fwd) or towards the left (RL_fwd). The second task was designed to distinguish forward from backward locomotion (FWD/BWD task). In that case, the stimulus shows a person walking towards the right, but either forward (LR_fwd)

(a) stimuli for the 6 speeds   (b) ankle trajectories

Fig. 2 Stimuli presented in the behavioral and the computational experiments. In (a), snapshots/body poses of the 6 walking speeds are depicted with the training speed framed. (b) Ankle trajectories for the same speeds: with increasing speed, step size increases as well as vertical displacements grow.

or backward (LR_bwd). The LR_bwd condition was generated by playing the LR_fwd video in reverse. The start frames of the movie stimuli were randomized across trials to avoid that the animals responded to a particular pose occurring at a particular time in the movie. Training was done only at the 4.2 km/h walking speed.

Substantial training was needed for our monkeys to learn FWD/BWD discriminations, while LR/RL discrimination was made more easily (cf. [4]). E.g., the number of trials required to reach 75% correct in a session for the LR/RL task was 1323 trials, while the same monkey required 37, 238 trials to achieve a similar performance level in the FWD/BWD task (similar trends were observed in the other two subjects). Nevertheless, all three monkeys reached behavioral proficiency at the end of the training sessions.

2.4 Generalization Test

Trained at one speed only, the monkeys were tested for generalization to other speeds in the two described tasks. This was realized by interleaving trials of the trained speed with trials from the other speeds in a 90:10 ratio. Moreover, in order to avoid associative learning on these new stimuli, we always rewarded the monkeys on these other speed stimuli (still correct responses on the trained speed were required to obtain a juice reward).

3   Computational Model

We recently developed a technique for the analysis of human behavior from unsupervised silhouette data [5]. In contrast to action recognition systems (e.g., [7]) where specific actions are trained, our approach models the training data in an unsupervised and generative manner in order to redetect familiar patterns at runtime. Unknown queries are rejected in the same spirit as in [8]. Our approach was initially developed to monitor the behavior of (elderly)

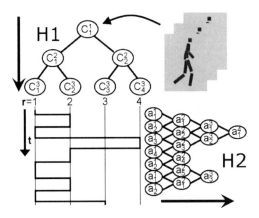

Fig. 3 Schematic overview of the developed computational model. From a set of data, *i.e.*, an image stream, our approach builds two hierarchies that encode the per frame appearance (snapshot, *H1*) and the actions (sequence, *H2*) in an unsupervised manner. *H1* is established from clustering the data in a hierarchical manner, *H2* analyzes how the cluster membership (the blue curve) evolves over time.

people in their homes, in this work we however apply it to the same data as used for the monkey behavioral study. An overview is depicted in Fig. 3 and described in the following.

3.1 Appearance Hierarchy (H1)

In a first hierarchical analysis stage, the per-frame appearance (snapshot) of the observed action is modeled. The training data are presented as normal video sequences of the *humanoid* stimuli. As input features we take binary segmented stimuli. These silhouettes are clustered (k-means) recursively in a top-down procedure, such that clusters are more specific on higher layers. This results in the structure *H1* as sketched in Fig. 3. The number of layers required, depends on the variability of the data. In this model, the root node cluster has to describe all training features, whereas leaf node clusters only contain similar data and are more precise. Data points that arrive at one leaf node cluster are given the corresponding symbol.

3.2 Action Hierarchy (H2)

In the action hierarchy, the sequence of symbols for subsequent frames in the training video data is encoded in a bottom-up process. Thus the evolution of the appearance over time is explored in *H2* as outlined in Fig. 3 (blue curve). If a change in the sequence of symbols appears, these symbols are combined to basic-level micro-actions, which are the low-level building blocks of *H2*. Inspired by [9], the combination of low-level micro-actions constitute higher-level micro-actions if they co-occur often enough. In this hierarchy a longer micro-action implies a more common action, as, again, such longer sequence is found to be common enough in the training data.

3.3   Analysis of Unseen Data

H1 and *H2* together provide a model of normal human behavior to which new data is compared at runtime. In *H1* a new frame's appearance is propagated as far as possible, seeking for the most specific cluster that still describes it well. Similarly, the observed history of appearances is matched in *H2* in order to evaluate how normal the observed action is. In [5] we show that this model can be used to track the (human) target, and that the model can be updated during runtime.

3.4   Generalization Test

We use the same stimuli for training and tested the same tasks with the computational model as for the monkeys (*cf.* Sec. 2). After training with 4.2 km/h walking stimuli, new stimuli are presented to the model with different walking speeds in order to test the generalization capacity.

Since the model is designed to cope with larger amounts of data, in which recurring patterns are detected, 6 repetitions of the same stimuli were used for training. The appearance hierarchy (*H1*) consists of 4 layers resulting in 8 leaf node clusters. Separate models were trained for LR_fwd, RL_fwd and LR_bwd. During testing, we use the LR_fwd and the RL_fwd models for the LR/RL task, whereas in the FWD/BWD task we apply the LR_fwd and the LR_bwd models.

In the test phase, each applied model delivers two output values how well each test frame matches *H1* and *H2* (assuming *H1* has validated the stimulus), respectively. The value for *H1* captures the appearance only by measuring the similarity to one of the leaf node cluster centers. Additionally, *H2* requires the correct motion and searches for a corresponding micro-action with maximal length. To finally achieve the output score (appearance score from *H1*, sequence score from *H2*) we combine the two models, each trained for one of the two conditions relevant for the task. They are evaluated at each frame and a likelihood ratio is calculated and averaged across the whole stimulus. If for example a stimulus walking from left to right is described well in LR_fwd, but not in RL_fwd, the resulting score is high. On the other hand, if both models perform equally well, no clear decision can be drawn and the score is close to 1 (chance level in the case of the computational model).

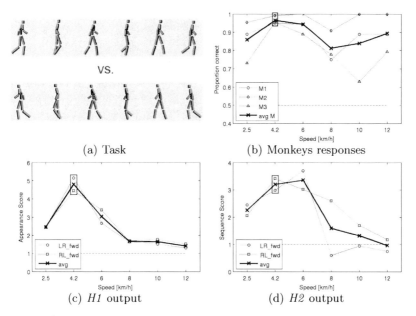

(a) Task (b) Monkeys responses

(c) *H1* output (d) *H2* output

Fig. 4 LR/RL task: Monkeys performances and model scores for 6 tested speeds.

4 Results and Discussion

In this section, the generalization performance for the different walking speeds is presented and compared between monkey behavior and computational model.

4.1 LR/RL Task

The results are depicted in Fig. 4, the monkeys responses are shown in panel (b), the appearance score (*H1*) in panel (c) and the sequence score (*H2*) in (d). Bold lines indicate the average results, dotted lines display individual performances for monkeys or different stimuli. Black boxes at 4.2 km/h point out the training speed. Chance level is marked with the dashed horizontal line.

In the behavioral study (Fig. 4(b)), categorization generalizes relatively well across the different walking and running speeds (binomial tests; $p < 0.05$ for 14 out of 15 generalization points). This suggests that the discrimination is based on spatial or motion cues that are common to the different speeds.

For the computational part, the results for *H2* (Fig. 4(d)) indicate a similar interpretation for slower walking speeds (2.5-6 km/h). In a more detailed analysis, we observe that for these speeds, the task can be solved already by only incorporating *H1*. Apparently, the appearances are distinctive enough.

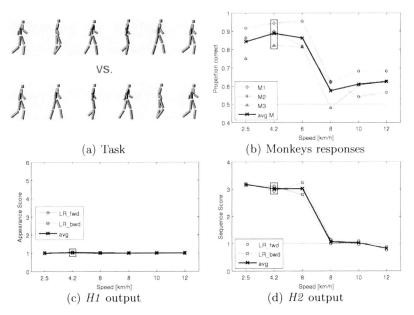

(a) Task (b) Monkeys responses

(c) *H1* output (d) *H2* output

Fig. 5 FWD/BWD task: Monkeys performances and model scores for 6 tested speeds.

For the running stimuli on the other hand, silhouettes are different, thus the performance in *H1* drops, which in turn drags down the *H2* scores.

4.2 FWD/BWD Task

The results for the FWD/BWD task are visualized in Fig. 5 in the same manner as for the previous task. The behavioral data from the speed-generalization tests show that the categorization is specific to walking: in each monkey, generalization is significant (binomial test: $p < 0.05$) for the walking, but not the running patterns. In fact, in each monkey there is an abrupt drop of the performance when the locomotion changes from walking to running.

Computational findings show that the evaluation of the appearance (body poses) only is not sufficient for solving this task (Fig. 5(c)). This is not surprising, since the appearances are the same for both stimuli. However, if their ordering is considered (Fig. 5(d)), the task is solvable for walking speeds, and the scores resemble those of the monkeys. At higher speeds, due to wrong appearance classification in *H1*, the sequence is not reliable in *H2* anymore.

The lack of significant transfer from the trained walking to running suggests that the animals learned a particular motion trajectory "template". Indeed, examination of the ankle trajectories (*cf.* Fig. 2(b)) reveals a relatively high similarity between those trajectories for the three walking speeds, which are in turn rather distinct from those of the three running patterns. This might also be a reason for the performance drop of the computational model.

4.3 General Discussion

At the behavioral level in our monkeys we noticed a clear qualitative difference in generalization performances across tasks. Whereas the monkeys were quite apt at discriminating other speeds not seen before in the LR/RL task, a clear step-wise function was observed in the FWD/BWD task. In the LR/RL, when confronted with other walking speeds, *i.e.*, 2.5 or 6 km/h, all three monkeys could correctly categorize these locomotions significantly higher than chance level. However this was not the case when confronted with locomotions at running speeds, again a trend present in all three monkeys.

The broader generalization observed in the LR/RL task compared to the FWD/BWD task shows that such motion cues are less specific. Alternatively, the monkeys might have used spatial features that are common to the walking and running *humanoids* that face in a particular direction. The fact that one could solve the LR/RL task quite simply by basing decisions on the presentation of just one frame could explain the observed (almost) perfect generalization. This is analogous to the first hierarchical analysis stage (*H1*), which works on the per-frame appearances of actions. However, at this stage, the model shows a slightly different pattern, performing quite robustly for the trained locomotion, with clear drop-offs already for the neighboring speeds. This is clearly due to overfitting of the model to the trained action. Monkeys have been exposed to other locomotion patterns before, in contrast to the computational model. When implementing the second hierarchical stage of the computational model (*H2*), which incorporates the evolution of the per-frame appearances over time, the model's performance resembles the monkey's performances more closely, especially for the FWD/BWD task. In summary, we see that monkeys have the capability to generalize well for simple tasks where snapshot information is sufficient. This might be due to prior knowledge based on different functional features, which is so far not included in the computational model at all.

5 Conclusions

In this work, we compared findings from behavioral studies to a particular, biologically inspired computer vision algorithm. The most important outcome is that a two stage computational system can, to some extent, reproduce monkey responses. The algorithm however has not the same generalization capacities which suggests that monkeys integrate the training in a broader manner than the computer does.

References

1. Giese, M.A., Poggio, T.: Neural mechanisms for the recognition of biological movements. Nature Reviews, Neuroscience 4, 179–192 (2003)
2. Schindler, K., van Gool, L.: Action snippets: How many frames does human action recognition require? In: Proc. CVPR (2008)
3. Lange, J., Lappe, M.: A model of biological motion perception from con gural form cues. Journal of Neurosciences 26, 2894–2906 (2006)
4. Vangeneugden, J., Vancleef, K., Jaeggli, T., Gool, L.V., Vogels, R.: Discrimination of locomotion direction in impoverished displays of walkers by macaque monkeys. Journal of Vision 10, 1–19 (2010)
5. Nater, F., Grabner, H., Gool, L.V.: Exploiting simple hierarchies for unsupervised human behavior analysis. In: Proc. CVPR (2010)
6. Vangeneugden, J., Pollick, F., Vogels, R.: Functional di erentiation of macaque visual temporal cortical neurons using a parametric action space. Cerebral Cortex 19, 593–611 (2009)
7. Lv, F., Nevatia, R.: Single view human action recognition using key pose matching and viterbi path searching. In: Proc. CVPR (2007)
8. Boiman, O., Irani, M.: Detecting irregularities in images and in video. In: Proc. ICCV (2005)
9. Fidler, S., Berginc, G., Leonardis, A.: Hierarchical statistical learning of generic parts of object structure. In: Proc. CVPR (2006)

Author Index